T0296313

REMINISCENCES ABOUT A
GREAT PHYSICIST:

PAUL ADRIEN MAURICE DIRAC

Reminiscences about a great physicist:

Paul Adrien Maurice Dirac

Editors

Behram N. Kursunoglu

Center for Theoretical Studies, University of Miami

Eugene P. Wigner

Department of Physics, Princeton University

The right of the
University of Cambridge
to print and sell
all manner of books
was granted by
Henry VIII in 1534.
The University has printed
and published continuously
since 1584.

CAMBRIDGE UNIVERSITY PRESS

Cambridge   New York   Port Chester

Melbourne Sydney

CAMBRIDGE UNIVERSITY PRESS
Cambridge, New York, Melbourne, Madrid, Cape Town,
Singapore, São Paulo, Delhi, Tokyo, Mexico City

Cambridge University Press
The Edinburgh Building, Cambridge CB2 8RU, UK

Published in the United States of America by
Cambridge University Press, New York

www.cambridge.org
Information on this title: www.cambridge.org/9780521386883

First published 1987
First paperback edition 1990

A catalogue record for this publication is available from the British Library

Library of Congress Cataloguing in Publication data
Reminiscences about a great physicist.
1. Dirac, P. A. M. (Paul Adrien Maurice),
1902–84. 2. Physics–History. I. Kursunoglu,
Behram, 1928– . II. Wigner, Eugene Paul, 1902–
QC16.D57R461987 530´.092´4 86-33409

ISBN 978-0-521-34013-7 Hardback
ISBN 978-0-521-38688-3 Paperback

Contents

Contents

2 MORE SCIENTIFIC IDEAS

3 INFLUENCED AND INSPIRED BY ASSOCIATION

Contributors

Behram N. Kursunoglu
Center for Theoretical Studies, University of Miami

Eugene P. Wigner
Department of Physics, Princeton University

Richard Feynman
California Institute of Technology

Margit Dirac
Tallahassee, Florida

Sevda Kursunoglu
Center for Theoretical Studies, University of Miami

Joseph E. Lannutti
Department of Physics, Florida State University

Harish-Chandra (deceased)
The Institute for Advanced Study, Princeton

N. Kemmer
Department of Physics, University of Edinburgh, Scotland

Rudolf Peierls
Oxford, England

A. D. Krisch
The Harrison M. Randall Laboratory of Physics, University of Michigan

Henry King Stanford
Office of the President, University of Georgia

Contributors

R. H. Dalitz
Department of Theoretical Physics, University of Oxford, England

A. Pais
The Rockefeller University, New York

Laurie M. Brown
Department of Physics, Northwestern University

Helmut Rechenberg
Max-Planck Institut für Physik und Astrophysik, West Germany

William J. Marciano
Department of Physics, Brookhaven National Laboratory

Maurice Goldhaber
Department of Physics, Brookhaven National Laboratory

F. Hoyle
Cockley Moor, Dockray, England

P. A. M. Dirac (deceased)
Florida State University, Tallahassee, Florida

P. T. Matthews (deceased)
*Department of Applied Mathematics and Theoretical Physics,
University of Cambridge, England*

J. C. Polkinghorne
The Vicarage, Canterbury, England

Nevill F. Mott
Milton Keynes, England

Harry J. Lipkin
Chemical Engineering Division, Argonne National Laboratory

J. Weber
Department of Physics and Astronomy, University of Maryland

Willis E. Lamb, Jr.
Department of Physics, University of Arizona

Abdus Salam
International Centre for Theoretical Physics, Italy

A memorial to P. A. M. Dirac

Paul Adrien Maurice Dirac, born in Bristol, England, on August 8, 1902, died on October 20, 1984, just two months after his 82nd birthday. His loss ranks with the loss of Isaac Newton, James Clerk Maxwell, and Albert Einstein.

In 1969, the first J. Robert Oppenheimer Memorial Prize of the Center for Theoretical Studies was awarded to P. A. M. Dirac to honor the prize itself. He was introduced to the audience with the following words:

Professor Dirac was born in 1902 in Bristol, England. He was graduated from Bristol University and obtained a Ph.D. from the University of Cambridge. He decided to study theoretical physics at Cambridge and contributed tremendously to the then developing subject of quantum theory. Heisenberg matrix mechanics hardly could be regarded as a basis to formulate the laws of nature without Dirac's fundamental formulation. The quantum theory as we know it and apply it today is as constructed by Dirac.

The 1920's were very exciting years for the world of physics. Great discoveries were being made almost overnight. A very rich and rewarding path was opened with the advent of quantum theory. One of the Meccas for the leading physicists of that time was in Göttingen. Young Paul Adrien Maurice Dirac was one of the members of that group – one of the most versatile ones. By the age of 23, he had already written his classic papers which put the concepts of quantum theory on a sound mathematical basis. He reconciled the ideas of relativity with the ideas of quantum theory and invented the well-known relativistic wave equation predicting the existence of magnetic moment of electron and, hence, a new fact – the spin. He further predicted that every elementary particle with a spin $\frac{1}{2}h$ has a counterpart with the same mass but opposite electric charge – or that a particle has an antiparticle. It was a prophetic prediction. To the electron, the corresponding system is the positron, which was observed after the prediction in cosmic ray experiments by Anderson in 1932. Later, electron and positron pairs were actually produced in laboratory experiments. In the same way, the proton and neutron also should have their corresponding antiparticles. In view of their mass content being nearly 2000 times that of the electron, experimental observation had to await construction of a large accelerator to produce such particles by collisions of protons with nuclei. This was accomplished in 1955. His formulations of the quantum field theory, statistics of fields and particles, his work on gravitational waves, and also his prediction of magnetic monopoles stand as further monuments to his originality and deep understanding of natural phenomena.

The impact of Dirac's discoveries at the fundamental level has been far reaching, since even now our research in theoretical physics is guided by Dirac's ideas and his formulations. It is often customary when one has a new idea to ask if Dirac hasn't done something in this area. In most instances, it turns out that the subject matter has been dealt with by Dirac in depth, with clarity and originality.

His work, besides bringing him the Nobel Prize and many other honors and prizes, has been instrumental in the award of the Nobel Prize to many others for the work they have done on his ideas and in the paths opened by him. To cite a few examples: Willis E. Lamb, Jr., Julian Schwinger, Eugene P. Wigner, Richard Feynman, S. Tomonaga, C. D. Anderson, E. Segre, O. Chamberlain, and many others. What he achieved in his early 20's formed the basis on which the Nobel Foundation awarded him the Nobel Prize in 1933 at the age of 31.

Professor Dirac is known not only in physics but, also, in ordinary conversations, as a man who does not make trivial remarks. There is a definite deep and well-defined meaning in his every sentence, even though the sentences are not very frequent. He is endowed with all the virtues of a great man; he has no enmities, no dislike for any human being.

Professor Dirac is a free man in the true sense of the word. This makes him also very courageous. It is a great honor for us to participate tonight in these modest ceremonies to award Professor Dirac the J. Robert Oppenheimer Memorial Prize.*

The years 1949–52 in Cambridge were most rewarding and memorable experiences for a young man attending Professor Dirac's course, entitled 'Quantum Mechanics'. There were students from other disciplines, such as mathematics, biology, chemistry, and even philosophy. When I asked one of them, 'Do you understand what Professor Dirac is writing on the blackboard?' his answer was, 'No'. I said, 'Why then are you attending his classes so regularly?' He replied, 'Some of the things I understand, but most of the mathematical

---

* *The Development of Quantum Theory* (P. A. M. Dirac's Acceptance Speech, 1969). Gordon and Breach Science Publishers, New York.

language I do not understand. However, at least I shall be able to say one day that I was among the students sitting in Professor Dirac's class on quantum mechanics.' Cambridge colloquia for theoretical physics were held weekly in a room adjacent to Professor Dirac's office in the Arts Building, which was next door to the famous Cavendish Laboratory. There was only one comfortable chair in the colloquia room, and nobody could occupy it, since it was known as Dirac's chair, even if, once or twice in three years, Professor Dirac was absent from these theory colloquia. It was rather remarkable that Professor Dirac always slept in these colloquia, except when his name was mentioned by the speaker. You could then see the white part of his eyes, indicating that he was responding to the mention of his name. However, despite that, the simplest and the most intelligent questions were always asked by him. Sometimes, the answers seemed obvious, but careful analysis showed that the question itself was deep and could be the subject of another paper.

Like other graduate students in Cambridge, I visited him in his office occasionally to show him some of my communications on unified field theory with Albert Einstein and Erwin Schrödinger, and he would read them carefully and then ask me about the progress of my work. He was most interested in Einstein's latest approaches to the generalization of the relativistic theory of gravitation. In those days, Professor Dirac was very much interested in the possibility of reintroducing the concept of aether into physics. In this line of approach, he had assigned a problem to Marcelo Cini of Italy, who was visiting Cambridge for a year. Other students in those years were Paul T. Matthews, Richard Eden, S. F. Edwards, Abdus Salam, Suraj N. Gupta, and many others, including John Ward during his occasional visits. Nicholas Kemmer, who retired in 1981 from Edinburgh University as Professor of Physics (where he occupied the chair vacated by the late Max Born), was, as a Cambridge University Senior Lecturer, in charge of the colloquia at that time. Therefore, he chose the speakers in consultation with Professor Dirac.

During the prewar and postwar years, Professor Dirac was the most eminent and esteemed personality of physics, and Cambridge was the Mecca visited regularly by many other distinguished physicists. In those days, the visitors included (just to cite a few) Heisenberg, Born, Bhabba, and Gell-Mann (then a graduate student at MIT). When departing Cambridge, I went to Professor Dirac's office to bid him good-bye, and he asked me if I could calculate the fine structure constant from my version of Einstein's unified field theory. I answered, 'I don't know how.' I then asked him, 'Can you calculate the fine structure constant from your theory?' He answered, 'In the future, I might.' Then, with the arrogance of youth, I said to him, 'In the future I might, too!'

In 1968, Professor Dirac was in Trieste at Salam's Centre for Theoretical Physics. I saw him in a restaurant there, and my wife and I proceeded to his table, where he was sitting with Mrs. Dirac (sister of Eugene Wigner), whom members of the family and close friends always call Moncie, instead of her name, Margit. I asked Professor Dirac if he remembered me, and he said, 'Yes, Kursunoglu.' During the conversation, I asked if he and Mrs. Dirac would like to visit the Center for Theoretical Studies in Coral Gables for a while. He then accepted a one-week stay but not more than that, because he did not like the heat. Professor Dirac arrived in December, 1968, with Mrs. Dirac and stayed several months before departing for Cambridge. He was back again in 1970 and 1971 for similar durations at the Center for Theoretical Studies. In 1970, when Professor Eugene Wigner called me one Sunday from Princeton regarding his forthcoming visit to the Center for Theoretical Studies, I informed him that his 'famous brother-in-law' was inquiring about his time of arrival so that he could be met at the airport, since, otherwise, he would get lost! Wigner replied in his usual polite way, 'I came all the way from Hungary, and I did not get lost. Why should I get lost in the Miami airport?' I said, 'In that case, I shall relay your remarks to Paul.' When I relayed his reply, Paul Dirac quickly responded, 'When Wigner came from Hungary, he had a lot of time. But, when he comes to the airport next week, he won't have that much time.'

Simplicity and clarity were among Professor Dirac's greatest attributes. One day, going around a small lake in Matheson Hammock Park in Coral Gables, I counted seven birds and told him I saw seven birds, and he said, 'No, there are eight.' I asked him, 'Where is the eighth?' He said, 'One just dove.'

Margit Dirac was greatly devoted to Paul. She tried to protect him in matters of health and prevent intrusions by press people and others who had no reason to see him. At social affairs, most of the stories were told by Mrs. Dirac, and Professor Dirac either merely nodded his head in approval or smiled when she related past occurrences in their lives. On matters of beauty and physics, Dirac has always been ahead of everybody else. On one occasion during a dinner in Fort Lauderdale, I said, 'Paul, just the other day, some people asked about my religion. I indicated to them that *beauty* was my religion, such as beauty in arts, in sciences, in nature, and in woman. Sometimes, I tell them that Professor Dirac has the same religion. Do you think I made a mistake?' Dirac added immediately, 'I shouldn't think so,' and he then asked, 'You aren't going to advertise this, are you?' With tongue in cheek, I replied, 'Of course not!'

Dirac moved from Coral Gables to Tallahassee in June of 1971 to join the physics department of Florida State University. Besides his attraction to a smaller city and the University, he wanted to be close to his daughter, Mary, who lives in Tallahassee.

The Center's annual Orbis Scientiae meetings on high energy physics began in 1964. Professor Dirac first attended in 1969. In this series, he was the first speaker on an annual basis. The meeting in January, 1983, was dedicated to his 80th year and, as in all previous 14 Orbis Scientiae, he presented a paper. The 1983 meeting was the 20th and the last in the Orbis Scientiae series.

When the annual J. Robert Oppenheimer Memorial Prize was awarded during the Orbis Scientiae conference, each recipient received his medal and citation from the hands of P. A. M. Dirac.

A special Festschrift prepared for him under the authorship of some of his friends, which was to be published in 1984 with a contribution by him, entitled 'The Inadequacies of Quantum Field Theory', will now, most regrettably, appear as a memorial to him.

Very few men and women of learning, science, letters, and arts can inscribe their achievements on history. Paul Dirac is certainly one of the most esteemed ones. His name will live forever.

Behram N. Kursunoglu
University of Miami

Preface

This book was planned, originally, as a Festschrift for Dirac's 80th year. Unfortunately, nature has allowed us to produce it only as a memorial volume. Paul Adrien Maurice Dirac was a most creative physicist who made many revolutionary contributions to the development of theoretical physics. The present volume contains contributions by some of his friends describing his human side, along with his profound influence in the evolution of modern physical theories.

Dirac was known by his students and contemporaries as a great genius who spoke very little and whose sentences were irreducible, in the sense that the addition or subtraction of even one word could modify their fundamental meanings. His humor, like his theories of physics, was original, logical, and unique. One example is described in Abraham Pais' contribution to this volume, 'Playing with Equations, the Dirac Way', where Dirac's comment on the novel *Crime and Punishment* was that the author made a mistake when he described the sun as rising twice on the same day. When one of us (BNK) asked Professor Dirac how he felt about the probability of the existence of some kind of life in other parts of the universe, he replied, 'If we do not find life in the universe other than on earth, then I must believe in the existence of god.'

The simplicity and profound logic of his thinking were reflected in all of his physical theories. This was, especially, clear when he said, 'God used beautiful mathematics in creating the world.'

The editors gratefully acknowledge the help received from Linda Scott, Helga S. Billings, and Lisa B. Scott in the preparation and organization of this book.

Editors
Behram N. Kursunoglu
Eugene P. Wigner
June 16, 1986

Chronology

Paul Adrien Maurice Dirac

1902   Born in Bristol, England, 8 August, son of Charles Adrien Ladislav Dirac and Florence Hannah née Holten

1914–18 Merchant Venturer's College, Bristol

1918–23 University of Bristol

1921   B.Sc. in Engineering

1923–6 Research Student, University of Cambridge

1926   Ph.D.

1927   Fellow, St. John's College, Cambridge

1930   Elected Fellow, Royal Society

1931   Corresponding Member, Soviet Academy of Sciences

1932–69 Lucasian Professor of Mathematics, Cambridge

1933   Nobel Prize in Physics

1934–5, 46–8, 58–9 Member, Institute for Advanced Study, Princeton

1937   Married Margit Wigner

1939   Royal Medal, Royal Society; Hon. Member, Indian Academy of Science

1943   Hon. Member, Chinese Physical Society

1944   Hon. Member, Royal Irish Academy

1946   Hon. Fellow, Royal Society of Edinburgh

1947   Hon. Fellow, National Institute of Sciences of India

1948   Hon. Member, American Physical Society

1949   Foreign Associate, U.S. National Academy of Sciences

1950   Foreign Hon. Member, American Academy of Arts and Sciences

1952   Copley Medal, Royal Society

1961   Member, Pontifical Academy of Sciences

1968–72   Member, Center for Theoretical Studies, University of
          Miami
1969   Oppenheimer Prize, Center for Theoretical Studies,
          University of Miami
1972–84   Professor of Physics, Florida State University, Tallahassee
1973   Order of Merit
1984   Died 20 October

1

HUMAN SIDE

1

Thinking of my darling Paul

Margit Dirac

My brother Eugene used to spend his summer holidays in Budapest, staying with our parents. That was in the early 1930s. The von Neumanns did likewise, sometimes taking the same boat from New York.

It happened on one of these visits, that I was included in an afternoon outing, to one of those fashionable tea places, near Budapest. It was a pleasant, relaxed outing, when Marietta suddenly turned to me saying, 'You must come and stay with us in Princeton.' I was happily surprised, and utterly delighted. Marietta was the most charming, witty, sparkling woman I ever met.

We spent the latter part of that summer with Eugene, in our summer house, 20 kilometers north of Budapest. Eugene insisted, 'If you come to Princeton, you must stay with me. What would people say, if you did not stay with your brother?' I was not terribly thrilled with this idea. The von Neumanns had a lovely home, they kept up the lifestyle to which they were used in Budapest, while my brother liked to appear, and act, like a pauper.

We sailed in the fall; Eugene had a two-bedroom apartment, proudly boasting that he furnished it to the cost of under $25. It looked like it. We had a dining room, and kitchen, well equipped as is usual in the States, but alas, when I grew up in Hungary, girls were expected to study languages, music, history or art, German and French literature and go to school, where you were given heaps of homework. All this failed to make a scholar of me, but it also deprived me of the slightest inclination towards cooking. Kitchens were out of bounds, not only for lack of time, but you were not supposed to risk smelling of onions. Consequently, we had to go out for our meals. My brother was used to that, and there were many charming places in, and near, Princeton.

It was soon after my arrival; we were having lunch at one of these restaurants, when a tall, slender young man entered the dining room, looked at Eugene and greeted him. He looked lost, and sad. I asked who he was, still standing undecided and none too happy looking. I was told, he was an English physicist, whom Eugene knew in Göttingen, where they used to have their meals together. 'He does not like to eat alone.' 'So why don't you ask him to join us?' This was how I met Paul Dirac.

That was the fall of 1934. The Institute for Advanced Studies had no building of its own as yet. Its members, like Einstein, von Neumann, and Dirac as a visiting member, had adjoining rooms in a large University building, called Fine Hall. I remember so well: to the left was Einstein's room, in the middle Eugene's and to the right of him, Dirac's.

Dirac had his sabbatical leave from Cambridge University. He came to the U.S.A. via Russia, where he visited some of his Russian friends. At one of our meals, he spoke about having to go up to New York, to buy himself a winter coat. He had left his in Russia – one of his great friends there had none. The friend was Tamm, the year 1934 when shortages in Russia were many and severe. He obviously dreaded a shopping day, while I was, and still am, an enthusiastic shopper.

Eugene was truly angelic with me. I had never felt this free in my life, I was 29, and for the first time, there was no pressure to get me down. So, I felt bold enough to suggest going up with him, to help him buy one. The suggestion was accepted at once. The day was set, I was to meet him in Fine Hall to drive to the station. Arriving first, having time to look round, I saw what looked like an airmail letter in Dirac's pigeon-hole. I took it, and put it in my handbag. In my excitement of going to New York, I forgot all about the letter. Only in the subway, having to open my handbag for a hankie, did I remember it, and handed it to him.

Dirac opened it and, as he was reading it, he seemed more and more concerned. When he finished reading, he turned to me saying that it was a letter from Cambridge; a great friend of his, a Russian professor at Cambridge University, who went every summer to Russia, to help with physics, was not being allowed to return to England. His wife wrote, saying that she was leaving England to join her husband with her two young sons. It was THE letter that Anna Kapitza wrote, and of course it was all very upsetting and sad. The Kapitzas were among the very few with whom Dirac had a great and affectionate friendship. Kapitza never took British nationality, and all his friends used to warn him of the possibility that the Russians one day would not allow him to return to England. This had now happened.

It was my first glimpse of New York shops. Paul had visited the U.S.A. on several previous occasions. He chose Lord and Taylor to find a winter coat. The superb coat, which served him to the end of his life, was soon bought. The salesman whispered to me, 'Don't you want to buy him a suit also?' The one he wore did look in dire need of replacement. In '34, Hungarians were not allowed to leave the country with more than a few pengös. Eugene supplied me with all I wanted and I kept a careful note, paying him back on his next visit to Hungary. I had to smile, and shake my head.

We continued having our meals together, and Paul's long silences became rarer. He spoke to me about his difficult, I should say very difficult, childhood. I told him about mine, which also left some sad memories and about my unhappy marriage. His domineering father made it a rule, to be spoken to only in French. Often he had to stay silent, because he was unable to express his needs in French. Having been forced to remain silent may have been the traumatic experience that made him a very silent man for life. His father died in '35; Paul was in Russia to watch the eclipse of the sun. As soon as he heard of the seriousness of the illness, he flew home to Bristol. It was too late. The first letter he wrote to me after his father's death was to say, 'I feel much freer now.' Anyone who knew my Paul will realise, for a sensitive gentle person to feel and write this way, what the past must have been like.

His father's family originated in France, from the village Dirac, with the Chateau Dirac. The family left France during the Napoleonic wars, and settled in Geneva, Switzerland. His father left his family abruptly, because of the severity of his parents. He went to Bristol, England, where he met Paul's mother, the daughter of a ship's captain. She worked in the University Library, and there they met. His mother was one of four beautiful daughters, and two brothers. It is the irony which only life can produce that Paul suffered severely from his father, who had the same difficulties with his family. Paul, although not a domineering father, kept himself too aloof from his children. That history repeats itself is only too true in the Dirac family.

Paul loved travelling; when term was over, he was off on the first day. In the summer of 1935, he came to Budapest, via Russia, Mongolia, China, and the Orient. Still in the same suit. His two suitcases were crammed with presents he had got in the East. He told us, how difficult it had been when shown beautiful things and agreeing their beauty, to be then presented with them.

Summers in Hungary used to be hot and lovely. My parents, with whom Paul stayed, lived opposite the Count Bathányi Park, where building was forbidden, on the hilly side, called Buda, a few minutes' walk from the Hotel Gellért, with its famous, unique artificial swimming pool. I lived somewhat higher up the hill, called Gellért mountain. I did not live in my own house, but had an apartment in Archduke Frederick's house, overlooking the town, with a fairy-tale-like view. So, down I went, to fetch Paul, for a swim. The people frequenting the Gellért were sophisticated, so, when I was shown the swim-suit, I knew Paul was in trouble – a huge hole in the seat. This might cause no alarm in Russia, but here? Even a policeman might be called. It was a battle, before another one was bought. Vanity was not one of Paul's vices.

Christmas 1935, I took my two children and Hilda, their Austrian governess (languages), to Mariacell, near Vienna. Paul came to visit us there. He used to go off for long walks; he knew no fatigue, meals were unimportant to him, but not to me. This incompatibility stayed with us through life.

Mariacell is superbly beautiful, especially in winter. It was a treat to walk among the lovely trees – wonderful views. I often accompanied Paul, but usually regretted it. His enduring capacity would have been too much for most mortals.

In 1936, my brother wrote about his plan to marry a young American physicist he had met in Madison, Wisconsin. He asked that one of us should come and meet her.

It was an enchanting thought to visit the U.S.A. once more. I volunteered with enthusiasm, and soon started out on the journey. I sailed from Southampton, on the Queen Mary; it happened to be her maiden voyage. It was a long train ride from Budapest to London, where I stayed for two days in the Grosvenor House Hotel. Paul invited me to lunch in Cambridge. He fetched me from Cambridge station, and took me to our hosts, Bishop Whitehead and his wife – the parents of Henry Whitehead, the mathematician. Mrs. Whitehead was Paul's friend and confidante, I think a sort of mother confessor. Years later Paul did tell me that he sought Mrs. Whitehead's opinion of me. She liked me, and again many years later Paul told me Mrs. Whitehead thought I had a childlike simplicity. If this was so I have been the right one, as we shared this so-called simplicity throughout our lives.

Next day Paul came to London to fetch me, and take me down to Southampton in his car. I boarded the Queen Mary only to be told that my visa for the U.S.A. had the wrong date, it was ten years too early. An important-looking young officer drove me to a big house and took me into an office where we spent, what seemed to me, long hours. To my question as to whether I would miss the boat, the young officer assured me, 'It cannot sail, before I am back on it.' It happened to be the U.S. Consul, who took me under his wing.

Paul stayed in Southampton, and watched the boat sail. He wrote saying what a magnificent spectacle it was.

In New York, Eugene was waiting for me at the pier. I stayed a few days in Princeton, then took the train to Madison, and there I stayed with the Gregory Breits. I met Amelia Franck, whom Eugene married a few weeks later. She was a tiny, lovely, dear girl in every way.

Soon I sailed back to Europe, arriving in Southampton, where Paul was waiting for me, and drove me up to London. Then and there, we decided to get married. Paul already had his mother up from Bristol. She was a kind, dear lady, and we always got on well. During the war, when Bristol had so many bombs, she was hurt taking her usual walk. Paul brought her to Cambridge, and she stayed with us until she died in 1942.

Little did Paul dread more than reporters. Whenever at all possible, he avoided them, and there are stories of how he managed that in Stockholm in 1933, when he got his Nobel Prize. His mother was with him, and he managed beautifully to avoid them most of the time.

It was a very quiet, small wedding. The Blacketts, who were our witnesses, Paul's mother, and we two, had lunch and then off we drove to Brighton, for two weeks. It was during the Christmas vacation, so Paul could stay away from Cambridge. Our wedding was on the 2nd of January, 1937, and so started a very old-fashioned, Victorian marriage.

Later, during that same Christmas holiday, I travelled back to

Budapest, taking Paul's sister with me. In Budapest, Betty met a good friend of mine, who lived in Amsterdam. She eventually married him and, from then on, there were two Hungarians in the Dirac family.

During the summer of 1937, Paul took me to Russia, where we visited his dear, wonderful friend, Peter Kapitza. He became the hero of my life – the one great man who was brave enough to say 'No' to Stalin, refusing to participate in making the hydrogen bomb. He endured five years of house arrest, had to live in his dacha, winter and summer, outside Moscow, isolated, lonely going on with his work, where his only help was his wife Anna. He risked daily for all those years hour by hour being shot and killed – he never wavered. His love for humanity was great, his courage boundless. This unforgettable friend we both loved and cherished enormously. He died at the age of 86, a few months before Paul.

2

Dirac in Coral Gables

Sevda A. Kursunoglu

Professor and Mrs. Dirac first arrived in Coral Gables, Florida, in December 1968, which was the month of the High-Energy Physics Conference at the University of Miami's Center for Theoretical Studies.

My husband, Behram, who was the chairman of the conference, thought it would be a good idea to invite Professor Dirac to the University at the time of the conference, which was already very popular with physicists all over the world. Many physicists, young and old, came to Coral Gables to attend these conferences to talk physics, meet famous people, and enjoy Florida sunshine. An application form was sent from the Center for Theoretical Studies (CTS) to Professor Dirac in Cambridge for him to fill out, as was the usual procedure for members of the Center. Under 'Proposed Field of Work', he wrote, 'Quantum theory'. That form is framed and is displayed at the Center today.

The conference took place at the new Science Building that year. The participants were greeted with music played by a group of musicians from the University's School of Music. Professor Dirac arrived in time to listen to the music and gave the first lecture at 9.00 a.m. When he finished his lecture, he sat in a front row seat and attended the rest of the morning session. At 10.30 a.m., when it was time for the coffee break, he was surrounded by participants who were very excited to see him in person. People stood in line to talk to him or to just greet him. Among his friends in attendance were Maurice Goldhaber, Willis Lamb, Julian

9

1968 Orbis Scientiae at the Center for Theoretical Studies. Front row, left to right: P. A. M. Dirac, Sevda Kursunoglu, Maurice Gusman, Margit Dirac, Edward Tellar, Lars Onsager, Mrs Teller and Behram Kursunoglu.

CENTER FOR THEORETICAL STUDIES
UNIVERSITY OF MIAMI
CORAL GABLES, FLORIDA 33124

Application for Membership

Period of Membership *Dec 27, 1968* from *Dec 27 1968* to *April 8 1969*, inclusive

Name in Full *Paul Adrien Maurice Dirac*   Date of Application _____

Permanent Address *7 Cavendish Avenue Cambridge, England*   Citizenship *British*

Present Address *Sheraton Four Ambassador*   Place of Birth *Bristol England*

Academic Degrees (Please give dates and places received)   Date of Birth *8-8-1902*

B.S./A.B. _____   Marital Status *M*

M.S./M.A. _____   Number of Children _____

Ph.D. _____   Ages of Children _____

Other _____   Please indicate whether your wife and/or children will accompany you:
☒ Wife   ☐ Children

Proposed Field of Work *Quantum theory*

Former and Present Positions and Fellowships

Dates	Name	Place	Rank

Honors and Societies _____

Intended Research: Please submit with one copy of this application a brief outline of your current and intended research.
Publications: A representative list of publications should be attached.

Signature *P A M Dirac*

(Please return one copy of this form to the Center)

Schwinger, Edward Teller, and his brother-in-law, Eugene Wigner.

That year, Professor Wigner came with his wife, Mary, and his daughter, Martha. They were staying at the Key Colony Hotel on Key Biscayne, where Professor Dirac, Mrs. Dirac, Behram, and I went to visit them in the evening. Professor Wigner recited some poetry from memory in Hungarian, and he gave us a short summary in English.

At the end of the conference on Saturday, Professor Sklar, who was a Professor in the Physics Department, rented a glass-bottom boat and invited the Diracs, Wigners, some of the participants, and us, to go on a day's cruise with his family. Professor Dirac was very happy and seemed to enjoy this outing very much.

In the early 1970s, when Professor Dirac was a visiting professor at the CTS, they stayed in an apartment maintained by the University for visiting dignitaries at the Four Ambassadors Hotel, overlooking Biscayne Bay in downtown Miami. Mrs. Dirac made life very pleasant for Professor Dirac and everyone else with her gracious hospitality at afternoon teas and ladies' luncheons and by giving small dinner parties. Every day, Mrs. Dirac would drive him to the CTS at about 10.00 a.m., and he would spend all morning in his room on the 3rd floor of the Computing Center. People at the Center named the room 'Dirac's room'.

He went to lunch at the Student Union Cafeteria with Behram and with other members of the Center, picked his own meal, carried his own tray and sat at the dining room tables where the students had lunch. After lunch, he walked back by the lake near the Student Union and returned to his room. Sometimes, he went to the nearby Faculty Club for lunch with Behram. Sometimes they went by car and sometimes they walked, as it took only about twenty minutes. I asked him once if he ever got tired after lunch and wanted to take a nap. He said he did nap in his chair or put his head on his desk to sleep to refresh himself for the rest of the day. I asked him, 'When you are working, do you ever get stuck?' He said, 'All the time'.

Three times a week, at 3 o'clock in the afternoon, he gave a lecture on 'Quantum mechanics'. His lectures were well attended by University of Miami students from different departments, such as physics, mathematics, medicine, and other sciences.

One of the busiest highways in the United States, U.S.1, stretches from Maine to Key West, Florida, and borders the University of Miami campus on one side. Despite its traffic, after his lectures, he liked to walk back to the hotel along that highway. He would be back at the Four Ambassadors in about one hour. Sometimes, he would get lost among the back streets and along the railroad tracks, but he always got home safely. Mathematics professor Robert Kelley, who knew Professor Dirac, recalls that 'He loved to walk. The first day he spent at the CTS, he walked home (to the Four Ambassadors Hotel) up U.S.1, which is at least five miles.' He liked to go for a swim in the hotel's pool, after he returned. After his swim, he would walk around the pool, which was by the ocean, and then take the elevator to his apartment on the eighth floor.

Sometimes, when I was having tea with Mrs. Dirac up in their apartment, I would look down from the window and see him getting out of the pool, relaxed and content.

Every Sunday, the Diracs went for an outing in their car. Sometimes, Behram drove them places, and sometimes I went along with them to such sights as Fairchild Tropical Garden, the Parrot Jungle, the Orchid Jungle, the Monkey Jungle, or the Seaquarium. He liked the many rare exotic plants at Fairchild Garden, but was especially fond of the Seaquarium, where he watched the sharks in captivity and enjoyed the porpoise show very much, commenting, 'They do clever tricks'.

Once, when we were visiting the Monkey Jungle, it was time for the feeding of the animals. A giant gorilla helped himself to a spring onion and ate it with great delight. Professor Dirac was very much entertained by his behavior. That evening, I prepared dinner for the Diracs at our house. Since he liked my mushroom sauce, I always made it for him with chicken or beef. On that occasion, I also served spring onions as a salad decoration. Professor Dirac took one and said: 'I am going to have one because the gorilla seemed to enjoy his onion.' I used to serve green peas as vegetables, because I noticed that he liked them. One time, he showed me a chart from a newspaper clipping which showed that canned peas had more nourishment than frozen peas. Many years later, I learned from his daughter, Monica, that her father missed mushrooms so much during the war that he tried to cultivate them in their garden in Cambridge.

One of Professor Dirac's favorite outings on a nice Sunday was to be driven to Everglades National Park, which is about a one and a half hour drive from Coral Gables. He liked to walk on the wooden trail over the fresh water marsh, from where he could watch anhingas, herons, egrets, grebes, ducks, water snakes, and fish. In the lake full of water plants, anhinga would come up on the surface of the water and swallow a fish or a small snake, or hold it in its beak, and dive right back in the water. Professor Dirac commented on white herons wandering around in the wilderness and disappearing from sight among tall thick sawgrass. He would notice the anhingas nesting in the bushes by the lake.

There were alligators, large and small, lying on the banks of the lake sunning themselves. Once, a huge alligator was trying to cross the road while we were driving in the car, and Behram had to stop the car to let it pass. Professor Dirac enjoyed the whole affair.

Dr. Kelley, who is a professor of mathematics at the University of Miami and president of the local Audubon Society, went to Everglades National Park with Professor Dirac many times. He would set up his telescope, and Professor Dirac would look through it. Dr. Kelley would spot an interesting bird and lend his field glasses to Professor Dirac, explaining to him what was happening. After one of their outings, Professor Kelley invited him to his home for dinner. Professor Dirac used to tell me that he enjoyed his trips to the Everglades with Kelley. The following is Professor Kelley's recollection:

Professor Dirac was especially interested in hummingbirds. His first question to me was: 'Do hummingbirds migrate?' I explained that they did and how and where. He commented that he had only briefly seen one fly over his head one day on the beach at Cape Cod. I arranged for him to have lunch at my home (in February 1970) and was able to place him next to a shaving brush bush where he had over ten ruby-throated hummingbirds flying around him. He often mentioned this occasion to me.

Professor Dirac was invited to give a lecture at the Miami Museum of Science one evening. The lecture room was filled to capacity, and some people were standing up in the back to hear him talk: 'On the Birth of the Universe, The Big Bang Theory.' After the lecture, there was a reception in his honor at the Museum. Many prominent Miami citizens and community leaders were among the audience to listen to his lecture, meet him, shake his hand, and talk to him.

Dr. Henry King Stanford, our close friend, and then president of the University of Miami, was the first to entertain the Diracs. The Stanfords invited the Diracs to dinner at the president's residence, with many members of the University's Board of Trustees and prominent community leaders in attendance. Many of those prominent community leaders, as well as several members of the University faculty, entertained the Diracs in their homes, including Mr. James McLamore, chairman of the University's Board of Trustees and founder of the Burger King chain; Mr. Edward C. Fogg III, Farm Stores owner; Mr. Emil Gould, prominent real estate developer; Mr. R. Kirk Landon, American Bankers Insurance Group president; Professor Henry Field, anthropologist and visiting professor from Harvard; Mr. Joseph Smolian, department store owner; and Mr. George Wackenhut, founder of the large security company.

Mr. David Blumberg, a prominent member of the University's Board of Trustees, hosted a dinner party at his home in honor of Professor and Mrs. Dirac and for the distinguished physicists attending one of the Center's conferences. Many distinguished persons from the community gathered at their beautiful and elegant home in Gables Estates, which is on Biscayne Bay. I was seated next to Professor Dirac. Lee Blumberg, the hostess, was speaking with Professor Dirac and happened to inquire, 'What do you think of Kursunoglu's work? Do you work together?' Professor Dirac answered, 'No, we are rivals.' Behram said, 'I am infinitely honored.'

After the large parties, Professor and Mrs. Dirac would ask Behram and me about the various notables. We would talk late into the night about those gatherings and the many fine and impressive people who attended.

The *Galant Lady IV* was a 100-foot yacht which belonged to the late Lewis Rosenstiel, who was a great philanthropist and had a home on Miami Beach. His wife, Blanka, a very charming lady of aristocratic Polish heritage, invited Professor and Mrs. Dirac, Behram and me, and some of her other friends, for an outing on her yacht one Sunday. We all met at Palm Bay Yacht Club early in the morning, where we were greeted by the captain and crew and shown on board. Professor Dirac was very much at home with Blanka, her brother, and the rest of the guests. It was a very lively and elegant group. There were champagne toasts for Professor Dirac, and caviar and lavish hors d'oeuvres were served by the crew. Professor Dirac did not partake of the champagne, as he was a teetotaler who always had ginger ale without ice and did not eat much. He did not say much during these outings. He sat quietly, but, when the ladies went up to him and tried to socialize with him, he seemed to enjoy their attention and always responded with a gentle smile.

In 1972, Professor and Mrs. Dirac came in January again for the High-Energy Physics Conference. They rented a house in Coconut Grove on Tigertail Avenue, which is in walking distance from the University (about 45 minutes) and also near the ocean at South Bayshore Drive. The village atmosphere of the Coconut Grove area made it very suitable for his walks. Their daughter, Mary, came from Princeton to visit them and also attended the conference. During that visit, Mrs. Dirac invited my children to have tea with them at their home and to meet Mary. My two daughters, Sevil and Ayda, my son, Ismet, and the Diracs were having tea in the living room when the clock on the wall started chiming 4 o'clock. Professor Dirac was very annoyed and asked who played with the clock. Apparently, from their first day in the Coconut Grove house, the clock had chimed every fifteen minutes, on the hour, half hour, and quarter hour. Professor Dirac could not stop the chimes, so, when the clock finally wound down, he did not rewind it. Mary fixed it without knowing why it had stopped, and Professor Dirac was cross with her. Mary wanted to leave right away. The children were surprised and amused to see Professor Dirac so uncharacteristically angry and could not help laughing when they told me the story.

At the end of the conference, which lasted only four days, Mary was returning to Princeton. Behram was driving the Diracs to the airport on the new expressway. I-95 was a brand-new road to the airport. We were pressed for time, and Mary was woried that she might miss her plane. I said to Behram, 'The next exit is the airport.' Professor Dirac, in a stern tone, said to me, 'Let's not have back-seat drivers.' That was the only time I heard him speak to me in that tone of voice in the nearly 16 years I knew him. He was always very sweet and gentle with me.

We would go for long walks without saying a single word to each other, and we would both enjoy the walks. One time, he asked me to walk with him from Matheson Hammock Beach at Coral Gables to our house. It was such a long distance for me that I began to feel faint. He said, 'Let us go slowly, and you will be able to make it', and I did.

Professor Dirac enjoyed sailing very much. It was very convenient for him, since their Coconut Grove house was very near the Coral Reef Yacht Club. On weekends, he sailed with Paul Vaughan, a former research chemist for the Goodyear Company who made his fortune inventing a particular cellophane paper for industrial use. Paul Vaughan kept his boat, the *Flim Flam*, docked by his house on the bay in Gables Estates. He would bring the *Flim Flam* to Coral Reef Yacht Club in the morning, and Professor Dirac would board there for a day of sailing. He wore his sun hat and carried a big scarf in his pocket for the Miami ocean breeze, which is always very pleasant.

Before the 1969 epidemic destroyed thousands of coconut trees in Miami, we had many in our back yard. Professor Dirac was fascinated with them. Some weekends, we had cook-outs for him in our garden by the pool, and he would easily stretch his arm and pick a coconut from a tree. Even though they were still green, he would crack them with an axe, split them into halves, drain the white coconut milk through a strainer into a glass, and drink it right away. He said it was delicious. Then, he would scoop out the tender coconut meat with his pocket knife and eat some of it. He always ate very little and very slowly.

He loved to go to Crandon park on Key Biscayne, where there are thousands of coconut trees. The beach also attracted him with its lovely clean sand. Sometimes, I waded in the shallow water while he was swimming, but usually I sat on a bench with Mrs. Dirac and we watched him enjoy his swim. The big, white waves would come and hit him, but he was not afraid of them. I always made sure that a life guard was nearby. After his swims, he would go for a walk on the sandy beach and disappear from sight. Then, he would return, appearing in the distance, a tall, thin figure clad in a brief swimsuit. As we walked back to the car, he would sometimes pick up a brown coconut from the ground where it had fallen and carry it with him to the car.

One time, we stopped at Dr. Rhona's apartment on Key Biscayne, where the University of Miami's Rosenstiel School of Marine Sciences is situated. I do not recall how Dr. Rhona and Mrs. Dirac first met, but they became immediate friends with a common Hungarian heritage. Dr. Rhona worked as a research scientist at the Marine Laboratory. While the two of them spoke together in Hungarian, Professor Dirac would watch with amusement. He used to say to me, 'Moncie (as he called Mrs. Dirac) likes to talk Hungarian.' Although he did not speak Hungarian, it was of little matter, since, even when the conversation was in English, he used to fall asleep in his seat. Often, though, when he heard his name mentioned, he would open his eyes and look around, momentarily interrupting his nap.

We have a comfortable 'Florida Room' on one corner of our house. Behram uses it as his study. One day while Professor Dirac was visiting, he dozed off while sitting in one of the heavy, padded chairs he favored. When I came into the room and noticed his eyes open, I asked if he had had a nice rest. He said he could not sleep, because Behram was chewing gum and was making a lot of noise. Behram had quit smoking at that time and had started chewing gum as a substitute.

In those years, Behram was very strict with our young children. When the Diracs and our family would go out, the children would regularly ask to have pets. Behram resisted until, one evening after the children

17

had eaten dinner at the Dirac's house without us, Mrs. Dirac talked Behram into letting the children have a dog. Behram brought home a dog which we named Pasha, whom we all (especially Mrs. Dirac) came to love. Mrs. Dirac would bring a tall can of dog food and feed Pasha in the garden whenever they came to visit. One Saturday, Professor Dirac went with Behram and my son to an afternoon matinee to see the film *Swiss Family Robinson*. When the movie was over and they came out, they saw that the two of them were the only adults at the show.

On one occasion, I asked Professor Dirac what he remembered most about his father. He told me his father insisted only French be spoken at the dinner table. Since Professor Dirac did not feel comfortable speaking French, he did not speak at all.

He did not talk much of his father, but he was quite close to his mother. He often spoke of her, and he took her with him to Stockholm when he received his Nobel Prize. He began his account with their train ride to Sweden. Professor Dirac had one sleeper; his mother had another in the next compartment. On the morning of arrival, she overslept and did not hear the conductor knocking on the door early in the morning. When the train pulled up to the platform, the conductor came by her compartment. She dressed hurriedly, not having had time to pack, and let the conductor in. He started throwing her belongings out the window. Her clothes, hair brush, and comb were strewn on the platform as she bustled off the train. Professor Dirac recounted this tale in his gently smiling, almost abashed way. There were newspaper reporters looking for him, he added, and they were even more confused than his mother.

In Stockholm the next day, he attended to business, leaving his mother at the hotel. In the evening, when he went back to the hotel for his mother, she was gone. There was a dinner party that night, and everybody started getting worried. Eventually, everyone else gave up and left for the party, and he stayed behind to wait for her at the hotel. It grew very late, and he was about to call the police when, at midnight, she appeared happy and smiling, as if nothing had happened. She said she got on a boat to go sightseeing around the harbor, but it went far out to sea to another destination, and she had to stay on the boat until it returned to the harbor.

Another boat story he enjoyed telling was of his trip to Japan on a boat with Professor Heisenberg. When they arrived in Japan, reporters came on board where he was standing on deck beside Heisenberg. As one of the reporters approached them, Dirac turned his back and stepped aside. The reporter asked Heisenberg, 'Where is Paul Dirac?' Heisenberg shrugged his shoulders and said nothing. The reporter did not notice Professor Dirac. He interviewed Heisenberg and left. Dirac was happy that he successfully avoided them (he outsmarted them).

Despite his quiet ways, he was always full of surprises. One day, Mrs. Dirac asked me to go with them to Coconut Grove to an antique shop run by Mr. Campbell and his mother. Mrs. Dirac and I often went shopping together, and Professor Dirac liked to tease her. He would say, 'Did you go shopping with Sevda today? Tomorrow you will want to go back and return what you bought.' On this particular occasion, she wanted to make a major purchase, so she took her husband along. They looked at several paintings in the front room of the shop, before moving to the darker back room to see the larger paintings. Mrs. Dirac pointed to a big oil painting on the wall and said to me, 'You have better eyes, look at it. What do you think?' It was like a Toulouse-Lautrec painting; a lady wearing a long evening dress with yellow, orange, and red floral print.

I remarked, 'This is exactly like the dress I wore to the conference party last night!' Professor Dirac replied, 'Yours was low cut.' His answers were always short and to the point.

The Diracs, Behram, and I once went to the Coconut Grove sidewalk art show, where artists from all over the United States exhibit their work. Professor Dirac liked a painting of an old man looking behind some trees and wanted to buy it. Mrs. Dirac begged Behram to persuade him not to, because she thought it was a terrible picture. She said she could not bear to look at it. Somehow, we escaped without the painting.

Just as some places are known for their sunrises or sunsets, everyone seems to know about the 'moon over Miami'. Some nights, when they were visiting us, Professor Dirac would go out in front of the house with Behram and stand on the sidewalk away from the trees, looking for a comet in the sky.

One night, the four of us were going out and were standing by the car in front of the house. There was a full moon, and I noticed Professor Dirac looking around. He said to me, 'Those shoes would look nice on a dance floor'. I was wearing a pair of black tie evening shoes with thick

19

silver heels that I had purchased in London. I was really surprised. I said, 'Thank you'.

Professor Dirac was not an avid television viewer, but there were shows he would watch. On Sunday nights, if they happened to be at our house, he watched *The Masterpiece Theater* or *Forsythe Saga*. He liked Irene, Soames' wife; she was blonde and beautiful. When Neil Armstrong went to the moon, Professor Dirac watched it on television at our house. He sat in front of the television set all night, causing Mrs. Dirac to remark, 'Paul is glued to the television'. I recalled that I had met Colonel Frank Borman and mentioned to Professor Dirac that Mr. Gusman had introduced me to the former astronaut.

Maurice Gusman came to America from Russia when he was 14 years old. He became very successful, first as a pharmacist in New York and later as a banker in Cleveland, although he had no formal education. He was in his seventies, but he was still working at Southeast Bank, which he founded. Until his death, Mr. Gusman remained principal shareholder of this substantial institution.

I introduced Mr. Gusman to the Diracs, and they became friends. Mrs. Dirac once asked Mr. Gusman about his investments, as she was interested in learning about the market. Professor Dirac did not say much during these conversations. He used to say to me, 'Most people like to talk, and very few people like to listen.' Mr. Gusman said to Professor Dirac, 'I never sell stocks'. Professor Dirac replied, 'I never buy stocks', stressing every word with a soft voice and a gentle smile. He was very cautious in whatever he said or did. He liked to say, 'There is no fool like an old fool'.

Just after Mr. Gusman moved from the Imperial House on Miami Beach to a new apartment at Key Biscayne Towers at the end of the island near the lighthouse, he invited the Diracs and us to dinner. His butler-cook prepared a lovely meal of beef stroganoff, which was surpassed only by the ambience of the fine interior decoration and panoramic ocean view.

Mr. Gusman remarked during dinner that, as much as he enjoyed his new residence, he had many fond memories of his former apartment, including the visit for a week from his friends, Lord and Lady Snow.

I, too, recalled the Snows' visit, which had been sponsored by Mr. Gusman to enable us to hear lectures by Lord Snow. He gave two public lectures at the University on 'Taking Care of Excellence' and 'The

World Problem.' During a conversation I had with Lord Snow, he said that Paul Dirac was the greatest living Englishman of our times and most deserving of the 'Order of Merit' given Professor Dirac by the Queen. I was fascinated to see the Order of Merit among Professor Dirac's letters and medals of commendation, which he carried in a finely crafted case inside his attache case.

At the conclusion of that year's residence in Miami, he mentioned that he was very happy to return to Cambridge before it got really hot here. Our families had agreed to get together in England over the summer, so we called on the Diracs in Cambridge. Professor Dirac was to dine at St. John's College that night. The rule was that a fellow could not take his own wife to dinner at St. John's College; only another fellow could invite her. So, I went with Professor Dirac to St. John's to have dinner and to meet the other fellows. Professor Dirac wore his Cambridge gown, and we sat at a long table in the center of the big dining room. Professor Dirac sat in the first seat, and I sat next to him. There, I met Dr. MacLaurin, who was sitting across from Professor Dirac. Dr. MacLaurin had been in Istanbul in the 1930s. Professor Dirac said to him, 'Sevda comes from Cyprus.' I added, 'I come from the Turkish section of Cyprus. I first came to England as a student. Now, we are living in America, in Miami.' He was an interesting man, who spoke at length about his travels.

The younger fellows sat further down the table, all dressed in gowns. The candlelight dinner was served rather quickly, course after course: Roast beef, boiled onions in cream sauce, and roasted potatoes. I noticed the younger fellows sitting down from us were not talking, but eating as fast as they could. Professor Dirac explained later that they ate like this because supper was their main meal. It was still light when dinner ended, so Professor Dirac took me on a tour of the hall. We then went to Mrs. Besikowitch's house next door to meet Mrs. Dirac, who was visiting her.

On the Diracs' third visit to Miami, they rented a house with a swimming pool that was a 45-minute walk from the University. He used to swim in their pool every day. The house also was near Tahiti Beach, on the bay in Coral Gables. The beach had little open straw shelters called 'chickees', with picnic tables and seats. Professor Dirac liked to sit in the shade under the shelter when he was not in the water.

At one time, I had a very bad pain in my shoulder. I asked him if it might be from going into air conditioned places after being in the hot sun. He advised, 'When you go to sleep at night, cover yourself up, including your arms and hands. You must have uniform temperature of the body when you are sleeping'. This cured the problem, and I am eternally grateful for his advice.

After the conference in January, he told Behram that a position had been offered to him at Florida State University (F.S.U.) in Tallahassee. He liked the idea of living in a small university town, with its small roads, where he could walk more freely. He also preferred the cooler winter weather. His daughter, Mary, had moved to Tallahassee, so he decided to move there. He said he would come to the High-Energy Physics Conference, as usual, and he did.

Professor Kelley recalls, 'After he moved to Tallahassee, he returned to Miami for some time each year. When I was fortunate enough to see him again, he always liked to talk about his walks in Tallahassee and the turtles and birds he had seen. He was a keen observer of nature and always curious about what he saw. When he asked me questions, he always wanted short answers that he could think about. He did not like to be interrupted while thinking; he preferred to ask another question instead.'

In 1982, the conference was held at the Konover Hotel on Miami Beach. Not far from there is an excellent French restaurant called Cafe Chauveron, which I thought they may like. By that time, I felt free enough to socialize more and worry less about entertaining at home. I suggested to Behram that we try it, so, one evening, we went there with the Diracs for dinner. When I made the reservations, I specifically requested a well-lit table, as Professor Dirac did not like to eat in the dark. He wanted to order 'duck à l'orange', but it was served only for two. Mrs. Dirac declined but suggested that I share one with him, which I did. We had a most enjoyable meal and were pleased with the restaurant.

Behram was invited by the F.S.U. Physics Department to give a talk every year in December, and I went with him. In early years, Mrs. Dirac would pick us up at the airport, and we would meet Professor Dirac at their home. During coffee, she once remarked, 'Paul, look at me', and he looked up. She then said, 'Look at your eyes. They are popping out because Behram and Sevda have come to see you'.

One morning, Professor Dirac, his assistant, and Behram went canoeing. Upon their return to shore, the canoe tipped as they were getting out, and Behram got thoroughly soaked. He had to take off his shirt and trousers and hang them up to dry, while he lay in the sun like an alligator to dry. As the boats passed by, the people would stare. In the evening, when Professor Dirac told us the story, he was still laughing.

Mrs. Dirac always had sherry on the cocktail table. I got Professor Dirac's ginger ale ready without the ice and passed around the sherry for the other guests. Professor Dirac sat comfortably in a chair, looking out their living room window while having his drink. He would sit there peacefully, sometimes closing his eyes and sleeping quietly, even when others around him were talking or moving about. His movements were very gentle. He would get up slowly, walk slowly, take a step forward slowly, and speak slowly, as if he were counting every word he was using to express himself.

He had graceful hands with long, slender fingers covered with transparent white skin, like fine porcelain that could break easily. At the dinner table, there was harmony between his hands and the fine crystal glasses when he reached out to drink his water – always without ice.

In Tallahassee, we always stayed in a motel as guests of Florida State University and usually arranged our visits for Thursday and Friday. As soon as we arrived in Tallahassee on Thursday morning, Behram walked to the Physics Department with Professor Dirac after he exchanged greetings with Mrs. Dirac. It was business as usual all day long. They spent the morning with graduate students, and Behram gave a colloquium in the afternoon. I remember that, during one of our visits, Professor Dirac walked home for lunch and left Behram at the University with the physics people. Mrs. Dirac, Professor Dirac, and I sat down in their dining room to have lunch. Mrs. Dirac opened a can of salmon for Professor Dirac's lunch, because he liked canned salmon. After tasting it with a slice of white bread, he got up and went to the kitchen, returning with a bottle of brown vinegar. 'The salmon tastes better with a little bit of vinegar', he remarked, and poured some from the bottle onto the salmon. He seemed to enjoy his lunch very much. I asked him what kind of salmon it was, thinking that I could get it for him when they came to Miami. When we got up from the table, Professor Dirac went to the kitchen and brought the can of salmon to the dining room as I was clearing the table. He read the label to me: 'Alaska Sockeye Red Salmon'. Mrs. Dirac was very much concerned about what

23

he ate. She bought jumbo-size eggs for his breakfast, because he would never eat two regular eggs. He used to say, 'Moncie is always trying to feed me'.

Mrs. Dirac would say to Professor Dirac, 'Behram is 24 karat gold', and Professor Dirac would nod. But, sometimes, when she would get cross with Behram, she would say, 'You are copper now'.

One day, when Behram and Professor Dirac were riding in the elevator, Behram said to Professor Dirac, 'Your tummy is sticking out, Paul', chiding that he was not as thin as Mrs. Dirac made him out to be. That evening, Professor Dirac did not want to eat as much as Mrs. Dirac offered him, because he said Behram told him that he was putting on weight. Professor Dirac believed that people should eat sparingly.

One time, when some Japanese physicists were to be in Tallahassee to see Professor Dirac, he asked Behram to stay on through the weekend. He said he would like to take Behram and me on a glass-bottom boat on the biggest freshwater lake in Wacula Springs, Florida. It was a very nice, sunny day. I sat on one side of Professor Dirac and Behram on the other side. He pointed out to us the interesting things as the boat cruised around the lake. There was a kind of underwater show with all kinds of fish swimming beneath us. It was very colorful and fascinating to watch through the boat's glass bottom. Professor Dirac enjoyed watching the different schools of fish perform. He also pointed out to us the alligators – huge ones and their babies, which were almost invisible in the landscape, sunning on the shores of the lake. I told him that I thought alligators existed only in Everglades National Park near Miami. I asked rhetorically if he had heard the expression, 'See you later, alligator', to which the response is, 'After 'while crocodile'.

In November 1977, Behram started a new conference series. The first International Scientific Forum on Energy was dedicated to Dirac's brother-in-law, Professor Eugene Wigner, on his birthday. Behram invited Professor Dirac to attend that first Energy Forum, which was held in Fort Lauderdale at the Sunrise Inn on Fort Lauderdale's beach.

A social highlight of the conference was a party given by Mrs. Rosemary Bernstein in her beautiful oceanfront apartment at Point of Americas on Sunday night before the conference started. Cocktails were at 7.00 p.m., and dinner was served at 8.00 p.m. in the clubhouse party room. Among those invited were Professor and Mrs. Dirac, Professor and Mrs. Bethe, Professor Hofstadter, Professor and Mrs. Teller, Dr.

and Mrs. Pierre Zaleski, the French nuclear attache from Washington, D.C., and Mr. and Mrs. Jean Couture and Dr. and Mrs. Georges Vendryes from Paris. There were eight courses served, each with a different wine, and sherbet between to clear the palate. The French chef Mrs. Bernstein hired especially for this affair created a new dish with red snapper, a local fish, in honor of Behram and his new conference. The party continued until 2.00 a.m. The Russian representatives were late arriving, so they missed the party. Unfortunately, Professor Wigner's wife was very ill, and they could not come. Professor Dirac enjoyed himself, but he was very upset that his brother-in-law and sister-in-law missed the party.

On Monday morning at 9.00 a.m., the conference started. The participants were serenaded by four violin players as they entered the conference room, which was facing the ocean on three sides with big glass windows.

Professor Dirac attended the morning sessions. During the lunch breaks, he walked across the road to the beach and went for a swim in the ocean. Then, he would sit on a bench and look at the water or go for a walk on the beach. He would wear a pale yellow beach robe and his sun hat. Sometimes, Mrs. Dirac and I would go over and join him.

One day, while we were walking on the beach, he told me that he would teach me a saying: 'It is easy, if you remember the symmetry. Watch the symmetry'. He went on, 'When a man says yes, he means perhaps; when he says perhaps, he means no; when he says no, he is no diplomat. When a lady says no, she means perhaps; when she says perhaps, she means yes; when she says yes, she is no lady'. With a couple of repetitions, I learned it, and he was pleased.

The following day, Behram arranged a picnic for the conference participants at Birch Park, across the street from the Sunrise Inn. The hotel prepared box lunches, and I went to the picnic with Professor and Mrs. Dirac. Enjoying the picnic with us were Professor Basov and his wife, who were attending the conference from Russia with two gentlemen accompanying them, Professor and Mrs. Bethe, and Professor and Mrs. Teller. The presence of Professor Dirac at the picnic made it an event for the physicists in our group. This was a free afternoon for the participants. I went for a walk with some of our guests and then walked back to the hotel with Professor and Mrs. Dirac.

Professor Dirac liked the location and the Sunrise Inn so much that

Behram booked the same hotel for the High-Energy Physics Conference in January 1981. This was an exception, because this conference was always held in Coral Gables.

The conference party was in the ballroom of the Surf Club in Surfside, near Miami Beach. The orchestra started playing and, somehow, Professor Dirac was asked to start the dance with me. When I got up and walked to the dance floor with him, I did not know which way to turn. He said to me, 'This way, I have to lead'. The party was a great success and so was the High-Energy Physics Conference.

I frequently asked Professor Dirac's advice. At that time, I was one of the women who went back to school to continue their education. I had finished my B.A. degree in mathematics, and, with Professor Dirac's encouragement, I was working toward a master's degree, also in mathematics. Sometimes it was very hard to pursue my studies with my socially demanding lifestyle. Since we live nearby, I often liked to walk to the campus to use the library or study in my office. I found my studies very exciting, and I enjoyed spending my free time learning higher mathematics. At the same time, I was afraid of what others would be thinking about my behavior. So, one day, I asked Professor Dirac about it. He said to me, 'You come first. Think about yourself first. If nobody gets hurt, do it'.

Except once or twice when they were out of the country, Professor and Mrs. Dirac came from Tallahassee to Coral Gables every year, until January 1983, which was the 20th anniversary of the High-Energy Physics Conference. That conference was the last in that series and was dedicated to Professor Dirac's 80th birthday, which was the previous August. He did not like the idea that some people started celebrating his 80th birthday prematurely. The meeting took place at the Four Ambassadors Hotel, where he stayed the first year he was in Miami. It was just by coincidence that the circle was completed.

One of his students from Cambridge, Dr. Harish-Chandra, came to Miami from The Institute for Advanced Study in Princeton. He was accompanied by his wife, Lili. They, too, stayed at the Four Ambassadors. It was like old times, when Harish-Chandra and Behram were students in Cambridge. We went upstairs to Professor Dirac's apartment one day, and he was very happy to have Behram and Harish-Chandra with him. He felt sleepy, so Mrs. Dirac persuaded him to lie down on the sofa. Everybody had to leave, and I stayed in the apartment to see that he woke in time to go down for an interview which was scheduled for 3.00 p.m.

In the evening, Mr. and Mrs. Miguel Orlandini hosted a dinner reception for the conference participants at the former estate of Alcoa founder Arthur Vining Davis. The Orlandinis were the new owners of the estate that Davis named 'Journey's End'. Since they own Bolivian tin mines, when Behram introduced the host and hostess, he said, 'Tin can be converted into gold very easily'. Professor Dirac came on the bus from the Four Ambassadors with the other participants. Mrs. Dirac stayed home and had dinner with the Harish-Chandras, as he had a heart problem and did not want to overdo it.

That evening, many beautiful ladies wanted to have their pictures taken with Professor Dirac for the newspapers. He was in excellent form, and I was fortunate to have the opportunity to enjoy that evening with him. He had a very good dinner, and, at the end of the party, he said goodbye and went back to the hotel on the bus with the participants. As Behram said about Professor Dirac, 'By honoring Professor Dirac for his 80th birthday, we are honoring the party of the 20th High-Energy Physics Conference with his presence.'

In May 1984, Behram was invited to Tallahassee by Professor Lannutti, an experimental physicist and associate dean of arts and sciences at Florida State University, to give a lecture on 'Humanization of Nuclear Issues.' This was an evening lecture, at 8.00 p.m. on May 20th, for FSU students and the public.

We arrived in Tallahassee in the morning and took a taxi to the motel. When we called her, Mrs. Dirac picked us up and drove us to their home, where Professor Dirac was waiting for us. He was very happy to see Behram, as usual. After they exchanged greetings, Behram went to the University, and I stayed at home with Professor and Mrs. Dirac. The three of us had a quiet lunch at home.

Professor Lannutti dropped Behram at the Diracs' home later in the afternoon. In the evening, Professor and Mrs. Dirac, Behram, and I went out to a restaurant for dinner. We finished dinner in time to go to the lecture hall on the University campus. Professor Dirac sat in the front row, and Mrs. Dirac and I sat in the back with some friends who came to hear Behram speak. The lecture went well, and there was an active question time afterwards. The questions and answers went on until 10.30 p.m. At the end of the lecture, the Diracs dropped us off at the motel, and Mrs. Dirac drove home with Professor Dirac.

The following day, Behram and I spent the day with Professor and Mrs. Dirac at their home. In the afternoon, when we were ready to leave Tallahassee at the end of our visit, Behram went for a walk with Professor Dirac in the churchyard across the street from their house. Mrs. Dirac insisted that she take us to the airport in her car which was parked in their garage.

Professor Dirac came out throught the garage to see us off. Mrs. Dirac was driving. As we pulled out in the car, he stood in front of the house, straight and tall, smiling, his hand up waving goodbye to us.

Behram told me at the airport, while we were waiting for our flight, that this was the last time Professor Dirac would say goodbye to him. He said somehow he felt it. I was not aware of it. I thought I would see him again. I did not want to believe Behram, but, unfortunately, in five months, on October 20, 1984, the whole world knew that we had lost a dear friend.

3

Recollections of Paul Dirac at Florida State University

Joseph E. Lannutti
Florida State University

I cannot presume to describe the impact of Paul Dirac on Florida State University. However, I can give some personal recollections of Paul as a close friend and colleague during his years with us.

Although my knowledge and appreciation of Paul Dirac began in the childhood of my career in physics, I had never seen or heard him speak until the first time he came to visit here at Florida State in January of 1970. If I remember correctly, someone had met him at one of Behram Kursunoglu's conferences in Miami and invited him to visit.

I remember being emotionally moved by that first colloquium. Having studied and worked in physics for many years, the name Dirac was a basic constituent, an integral part of the warp and woof of my understanding of physics. Dirac was not a person. Dirac was an equation, a theory of anti-matter, a delta function, a monopole or a kind of statistics. He was only a name in textbooks and history – in association with other great names such as Einstein, Schroedinger, Heisenberg and Bohr. But, now, he was here, in person! Describing his work and his interactions with our other heros in textbooks!

My experience was not unusual. I have learned that it had occurred with almost everyone at Dirac lectures – always given to capacity crowds.

Having NSF Science Development Funds budgeted for visiting

P. A. M. Dirac, January 1972.

eminent professors, we invited him to come back for the fall semester. He agreed! We were elated! He spent the fall of 1970 with us.

About a week before the Diracs were to leave, at a Christmas party at my home, Margit Dirac took me aside to say that Paul liked it here. She asked if it would be possible to arrange a continuing appointment. I calmly responded that I would investigate. Of course, what I couldn't say was, that this was what we were hoping for all along!

Within two days, after a rapid sequence of meetings with Norman

Heydenberg (our department chairman), Bob Lawton (Dean of Arts and Sciences), Paul Craig (Vice President for Academic Affairs), Stan Marshall (President), and Bob Mautz (Chancellor of the State University System), Paul was given an offer and he accepted.

I especially remember that, since Bob Lawton was a Shakespeare scholar, he became more enthusiastic when I asked how he would respond if Shakespeare had asked to join our faculty!

The next 14 years are a kaleidoscope of memories and I will recite only a few personal ones:

Lunch on the 7th floor with Paul, Steve Edwards, and Bill Moulton. Paul walking to work every day.

Getting him on our High Energy Physics contract with the U.S. Department of Energy to support him with a research associate and a graduate student.

Comment by our DOE contract monitor, Bernard Hildebrand, concerning the need for peer review of the research proposal: 'Who could be a peer to Dirac? Whose opinion could he accept, if contrary?'

His refusal to accept an honorary degree from FSU, since he had consistently refused every other university in the world and he did not want to insult them.

His contention at a Mathematics Department seminar that the mathematics that applies in physics must be beautiful, since it was the form selected by God.

The female undergraduate student who came to his office and handed him a rose, commenting only that it was to respect and honor him, and then leaving without introduction.

My first non-family visitors after my heart surgery were Paul and Margit and Paul's strong recommendation that I do more walking for exercise.

Then, for more than a year, walking with Paul 4-to-10 miles on the forest fire trails of the Appalachicola National Forest each Saturday and Sunday – Lost Lake, Dog Lake, Silver Lake, behind the Aenon Church, around Lake Bradford, etc. Contemplating nature from a lakeside.

Recounting his experience as an air-raid warden during World War II.

Paul being prompted by Margit to tell jokes at teas in the Dirac living room.

31

Paul avoiding medical doctors. Until an introduction to Hank and
Henrietta Watt began a lasting friendship.

Paul's smile and sparkle whenever I visited him in his office.

His desk full of letters from people seeking his opinion. Once
complaining that people should have the courage and commitment to
proceed with their ideas themselves.

Surprise of an Encyclopedia Britannica interviewer when Paul said that
the Physics Departments at Cambridge and Florida State were about
the same quality.

Recounting his arguments at a meeting with the Governor that North
Florida water resources should not be shipped south.

His greatest fear was loss of intellectual prowess. But he continued to be
creative to the end, dictating a contribution for a Kapitza memorial
volume to his Research Associate, Leopold Halpern, a few days before
he died.

Even with a severe case of flu, because he had promised, he traveled to
Fermilab near Chicago and stood alone on stage in front of an
auditorium full of physicists and talked continuously for two hours
without rest, reciting historical anecdotes for a conference on the
history of physics.

His unusual and unique agreement to let us use his name on the main
road for Innovation Park. Evidence of his strong support for his
adopted community! He very much wanted to help Innovation Park
and Florida State.

Although he was one of the Fathers of Quantum Mechanics, he
contended that it was wrong, and young physicists should not accept
it so readily without question.

His grand programs continued to the end. The gravitational strength is
not constant, he said, and there are observable cosmological
consequences. The pathological representations of the Lorentz group
may hold clues for a new quantum mechanics, he said.

This was how I knew Paul A. M. Dirac. World leader in science.
Profoundly influential in the course of science and history.
Acquaintance of Einstein, the Queen, and the Pope. Known and
respected world-wide. Honored by academies of science throughout the

world. But, a modest and gentle human being. Probably Tallahassee's greatest resident ever.

An intellectual giant who lived and worked for us for almost 14 years. Uplifting and honoring us. His spirit lives on through his friends and his intellectual legacies.

4

My association with Professor Dirac

Harish-Chandra
Institute of Advanced Study, Princeton

As a young undergraduate in 1940, I came across a copy of Dirac's book *The Principles of Quantum Mechanics* – the first edition of 1930 – in the library of Allahabad University in India and was immediately fascinated by it. The exposition was so lucid and elegant that it gave me the illusion of having understood most of it and prompted in me a strong desire to devote my life to theoretical physics. K. S. Krishnan, who was appointed Professor of Physics at Allahabad two years later, encouraged me in every possible way and on the basis of his recommendation, Bhabha accepted me as a research student at Bangalore in 1943. At that time, Bhabha was mainly interested in two projects, the classical theory of point particles and the theory of particles of higher spin. Both of these had their origins in papers by Dirac. I did some work in these areas, some of it jointly with Bhabha, and he sent it for publication to the Royal Society. During the war, the mail between India and England was very slow and, therefore, as a special favor, Bhabha requested Dirac to correct the proofs of the first two papers. This is how Dirac came to hear about me and, eventually, to accept me as a research student at Cambridge in 1945.

Although the war had ended, things were far from normal when I arrived in England. Cambridge was almost deserted, and Dirac had only two other research students: S. Shanmugadhasan from Ceylon (now Sri Lanka), whom I had met briefly in Bangalore, and Sonja Ashauer from

34

Brazil. As a thesis problem, Dirac asked me to investigate the irreducible representations of the Lorentz group. The finite-dimensional representations were all known but none of them, except the trial one, is unitary. Therefore, he wanted me to look for infinite-dimensional unitary representations. This is how I got started in group representations.

In Cambridge, my personal contacts with Dirac were rather infrequent. I went to his lectures but soon dropped out when I discovered that they were almost the same as his book. However, I did attend the weekly colloquium run by him. I found that he was very gentle and kind and yet rather aloof and distant. Therefore, I felt that I should not bother him too much and went to see him only about once each term. At my request, he communicated my paper on the Lorentz group to the Royal Society and another one to the *Physical Review*. He also got me admitted to the Institute of Princeton in 1947. But, beyond this, I had no way of knowing what he thought of my work. During my stay at Cambridge, I attended the lectures of J. E. Littlewood and Philip Hall and discovered the exciting world of mathematics. In those days before renormalization, quantum electrodynamics was plagued with divergent integrals, and I began to have an uneasy feeling that what I was doing had very little connection with physical reality. This drove me to look for comfort in the security of pure mathematics.

Dirac was also spending the academic year '47–'48 in Princeton, and, in fact, I was appointed to be his assistant. Soon after coming to Princeton, I became aware that my work on the Lorentz group was based on somewhat shaky arguments. I had naively manipulated unbounded operators without paying any attention to their domains of definition. I once complained to Dirac about the fact that my proofs were not rigorous and he replied, 'I am not interested in proofs but only in what nature does.' This remark confirmed my growing conviction that I did not have the mysterious sixth sense which one needs in order to succeed in physics, and I soon decided to move over to mathematics.

I have often pondered over the roles of knowledge or experience on the one hand and imagination or intuition on the other, in the process of discovery. I believe that there is a certain fundamental conflict between the two, and knowledge, by advocating caution, tends to inhibit the flight of imagination. Therefore, a certain naiveté, unburdened by conventional wisdom, can sometimes be a positive asset. I regard

35

Dirac's discovery of the relativistic equation of the electron is a shining example of such a case. The arguments which he gave for his derivation are very suggestive, but, on closer examination, they do not seem to be entirely convincing. He said that the equation should be of the first order. But, if he had had a more conventional grounding in differential equations, he would have realized that the Klein–Gordon equation, as Kemmer showed later, can also be written as a first order equation. The second argument that the probability density ought to be positive was more cogent. But then his equation landed him in terrible difficulties with the negative energy states, and it took a very bold stroke of imagination to rescue it from there. In fact, the proposed solution, although physically very daring and attractive, was mathematically somewhat ungainly and awkward. Nevertheless, within a few years, nature conferred her approval on Dirac's marvellous insight by the discovery of the positron.

I have found that reading Dirac's papers is a memorable experience. Even though they always deal with some major problem, they are usually quite short. Knowing the difficulty of the subject, the reader is all geared for fierce skirmishes on the way where he will have to slay dragons before he can reach the coveted treasure. But, as the journey unfolds, no dragons materialize, and all the obstacles keep melting away under the magic spell of the expert guide. There are no complicated and laborious computations, and the argument proceeds so naturally and smoothly that there is an illusion of effortlessness. Finally, on reaching the goal, the reader cannot help wondering what the fuss was about and where did the difficulty of the problem lie.

In Princeton, the Diracs invited me to their house many times, and I came to know the whole family quite well. In the spring of '48, they took me on a short trip with them to Ithaca where Mrs. Dirac's sister and parents had a farm. They all received me very warmly, and, as a result, I have felt a close personal bond with the Diracs ever since.

Although I cannot claim that my work had close links with that of Professor Dirac, it is true that, as I said earlier, my long infatuation with group representations came about from a suggestion of Dirac. Moreover, his example was always a powerful guiding influence in my life. The awe and reverence which I have felt for him for over 40 years is now mixed with a feeling of gratitude and affection; therefore, I am very happy to have this opportunity of paying homage to the profound originality of his thought and the purity and gentleness of his spirit.

5

What Paul Dirac meant in my life

Nicholas Kemmer
University of Edinburgh

The name Dirac meant something important to me well before I had taken my first steps into theoretical physics. I have a vivid memory of a day during my final year at school, when, in an end-of-term talk on the latest in the physics of atoms, my science master went to the blackboard and wrote down a list of names:

	Planck	Einstein	Bohr	
Born				de Broglie
Heisenberg				Schrödinger
Jordan				
Dirac				

I didn't retain much of what was said in that talk, except that, in some way, the word 'wave' was related to the two names on the right and the mysterious word 'matrix' to those on the left. Since my schooldays ended in the spring of 1930 and Dirac's first great paper had appeared only four years earlier,* my schoolmaster must have broken some kind of a record in the communication field.†

Why I know for certain that Dirac's name stood on that blackboard is

---

* Not his first publication, which I believe was in print before Dirac was 22!

† I think this is made only slightly less remarkable in the light of the knowledge that he had a strong personal interest in the career of his former pupil Pascual Jordan, in whose footsteps I was urged to follow.

that, for quite some years since that day, I carried in my mind the picture of a Pantheon of modern science inhabited by just those nine people. Of course, at that time, I knew nothing about Dirac the man, not even in what part of the world he was at home.

In this little offering in memory of Paul Dirac, I should like to relate how my knowledge of the man and his work gradually developed. I want to testify that, whatever criterion might be applied to revise, enlarge, or reduce the list of occupants of my Pantheon, whether then or now, the name Dirac is there to stay.

My first introduction to Dirac's work came in 1932, when I attended a brilliant first course on quantum mechanics given by Gregor Wentzel at Zürich. I still have the lecture notes I made at the time and see that, towards the end of the course, we were presented with a very clear account of Dirac's work on the quantisation of the radiation field[1]. I have no similar definite evidence of my first encounter with the Dirac equation for the relativistic electron[2], but that could not have been much later. By the end of 1933, when I was starting on my first research problem, that equation was central to my work. I also remember vividly how, in a Zürich bookshop, I picked up their sole copy of Dirac's *Quantum Mechanics* in its first edition. I have that copy before me now with 26th October 1934 entered as the date of my purchase. Anyone familiar only with later editions would be surprised to see that bras and kets do not figure in it and that other standard material associated with Dirac's name is not yet there, but he would certainly recognise the typical Dirac style; the clarity, directness, and self-containedness – perhaps 'purity' is the word I want – of the presentation is unmistakeable.

I had heard from Wentzel of the 'transformation theory' of quantum mechanics, which somehow unified the different possible approaches to the subject[3], but, even so, the bold, sweeping exposition of this unification I found in the book astonished and thrilled me. Let me dwell for a moment on just one feature of Dirac's approach – the 'delta function'. My earlier studies had been centred on mathematics, mainly under instruction from purists and formalists. Reading of the delta function, I felt I was treading on forbidden ground, to be avoided by all good little mathematicians. Soon, however, I was eager to accept what Dirac offered. How different things are now after 'distribution theory' has provided respectability for the Dirac approach. What a contrast there is, to give but one small example, between the painstaking method of contour integration, by which Wentzel derived the retarded potentials for us, and the few lines that were sufficient when, much later, I had to present the same results in my lectures!

At the start of my work for a doctorate, Wentzel handed me a reprint that was important for the work he was doing at the time. That same yellowing booklet is also with me now. It is probably the first reprint of a scientific paper that I ever possessed and is certainly the first of Dirac's original publications that I studied. For Dirac, it is an unusual paper because it reports on a collaboration with Fock and Podolsky, who are co-signatories[4]. It was published in the long defunct *Physikalische Zeitschrift der Sowjetunion*. Wentzel was interested in it during his efforts to eliminate the self-energy of the electron described by the Dirac equation (in its original form, with negative energy levels). Wentzel's[5] efforts were, of course, not successful,* but he was right to have chosen the Dirac–Fock–Podolsky paper for study. It is a remarkable one; if one looks at it now, one can see in it, with its many-time description for a many-electron system and its technique of canonical transformations, the ancestry of the ideas of Tomonaga and Schwinger that led into Quantum Electrodynamics as it is presented today.

As for my own early research, the Dirac equation was in the centre of all I tried to do, and even a little later, when I was more interested in boson than in fermion fields, many of the ideas that I helped to develop grew out of what had been first encountered with Dirac's electron–positron field.

The history of that field itself provides an example of how far beyond other people's horizons Dirac was able to see when he formulated new ideas. True, when he first published the idea[7] of the vacuum with all negative energy electron states filled, he was not ready to break out of the confines of the then known world of physics by postulating a new particle and saw the proton in the role of the electron's antiparticle (which promptly led Pauli to dismiss the whole picture). However, as we all know, Dirac triumphed in the end[8].

I need not go on to enumerate forward moves in theoretical physics

---

* However, the formulation of electrodynamics contained in Wentzel's work attracted Dirac's attention[6]. The construction he named 'Wentzel field' was taken from the papers published by Wentzel at the time, and Dirac's 'lambda-limiting process' was foreshadowed in Wentzel's work.

that started in Dirac's mind. I am sure that many will be mentioned by others joining in this tribute. One thing, however, needs to be stressed. There have been huge developments in the physics of the last half-century from which Dirac stood aside. Many experimental results were, I am sure, as unexpected to him as to the rest of the physics world. Many trends have gone in directions that seemed unsatisfactory to him, even though they grew from the base he created. Quite a few of Dirac's papers, though always original and full of surprises, were too far removed from fashionable thought to have attracted the attention that they deserved. However, the unique feature of his work was that it tended to do the reverse of becoming dated. While other people's unfashionable contributions tend to rest in peace between the covers of journals to be entirely forgotten in due course, Dirac's work tends to grow topical with age. This is surely true of what he said about the variation of physical constants with time[9] – and what about the monopole?[10] So I say to colleagues more active than I am these days: if you unearth an old Dirac idea that seems of no significance – watch out, it may be the theory of tomorrow!

I remarked at the start of this offering that, in due course, Paul Dirac, the man, became important in my life. Let me briefly tell the story. As soon as I found myself back in England (where I had spent ten years of childhood), I was eager to meet that hero of mine in person. Cambridge was easily accessible, and I had friends there; also, there were regular meetings of physicists in London attended by many Cambridge folk. However, I have no memory of seeing Dirac during that period and, for certain, there was no significant personal contact. In 1940, I was moved to Cambridge to do work of which I was sworn to say nothing except to co-workers in the group. My presence in Cambridge was known, however, so that I met many physicists on a purely social level, and in due course, I also helped the depleted Cavendish staff with some lecturing. I still did not meet Dirac in University surroundings, as far as I remember. Naturally, I acquired a mental picture of him as aloof and unapproachable. Then the dear friends in whose house I was living told me one day that I was included in an invitation to afternoon tea with the Diracs – just down the road. That day, I found myself in a beautiful garden in an atmosphere of warmth and informality with Paul Dirac and his daughters on the lawn or among (or even up) the great trees. True, in general conversation, our dear, gracious hostess contributed rather more than her illustrious husband, but I soon knew that adjectives like aloof or cold to describe him would be very far from true. Paul Dirac, on whatever theme, said what there was to be said in a minimum of words matched to a maximum of content. The clarity of his spoken words was the same as in his writings, but there was added to it an element of warmth and humanity that immediately put one at ease. That tea party was only one of a number of happy, private occasions.

My wartime stay in Cambridge only lasted for a few years, but it was my great fortune that my first peacetime job brought me back there, but under quite different conditions. I re-emerged fully integrated into University and College life. Also, after not much more than a year, I married; our happiness was added to by the warmth with which my new wife – and in due course, our children – were received by the Diracs.

Now, in addition, I had the privilege of working in the Faculty of which the Lucasian Professor was the most distinguished member. This was the Faculty of Mathematics which was able to attract virtually all the most gifted young mathematicians from English schools and many more from elsewhere. In general, Pure Mathematics was the most prestigious subject, but Dirac's annual course of lectures to what would have been the final undergraduate year in other universities inspired many of the ablest students to turn to theoretical physics. In the Cambridge system there was, however, an additional year that was comparable to a post-graduate year elsewhere. Within this, I was entrusted with a two term course called 'Nuclear Physics'. In it, I was supposed to cover all that was known in those days about nuclei, fields and particles. Inspired by what they had learned from Dirac, many students came to hear me, and I am sure no young lecturer anywhere at the time could have been addressing a more gifted assembly of would-be theoretical physicists. From among them, there came a steady stream of applicants to continue on Ph.D. work – with, of course, Dirac as first choice for a supervisor. Since, however, working with even one research student did not always fit into Dirac's plans, I was overwhelmed with applicants whom I soon found difficult to accept. (I nearly refused Abdus Salam!) Happily, after a few years, I was joined by colleagues with whom I could share this work (and I could also pass on some of the work to colleagues elsewhere, particularly to Rudolf Peierls at Birmingham). There is, thus, no doubt in my mind that the development of the postwar school of theoretical physics at Cambridge all began with the impact of Dirac's Quantum Mechanics Course.

Within our postgraduate group in those early postwar days, Dirac's personal influence was also felt strongly. He regularly attended the seminars in which we sought to keep abreast with the flood of advances in our subject. Through Dirac's interventions, which were, as always, short and crisp, he could often do more to clarify our thinking than hours of discussion without him.

I terminated my service to theoretical physics at Cambridge when I moved to Edinburgh in 1953. Inevitably, contacts with Paul Dirac became much rarer, but, I am happy to say, they did not cease, nor did our family friendship. I had the privilege to help honour him in Trieste when he reached his 70th year, and subsequently we in Edinburgh were honoured by a visit by him and his dear wife. Paul delivered a public lecture, and we had a happy family reunion. The lecture proved that Paul Dirac could still enthral an audience – I even had one letter from a non-physicist member of the audience describing the lecture as the best he had ever attended.

Now that Paul Dirac has passed on, I offer this as my very personal tribute and way of saying, 'Thank you for everything'.

References

1 P. A. M. Dirac, *Proc. Roy. Soc.* **114**, 243 (1927).
2 P. A. M. Dirac, *Proc. Roy. Soc.* **117**, 610 and **118**, 351 (1928).
3 P. A. M. Dirac, *Proc. Roy. Soc.* **113**, 621 (1927).
4 P. A. M. Dirac, V. A. Fock & B. Podolsky, *Phys. Zeitsch. d. Sow. Un.* **2**, 468 (1932).
5 G. Wentzel, *Zeit. f. Phys.* **86**, 479 (1933).
6 P. A. M. Dirac, *Proc. Roy. Soc.* **167A**, 148 (1938), and *Phil. Mag. 7th Ser.* **39**, 31 (1943).
7 P. A. M. Dirac, *Proc. Roy. Soc.* **126**, 260 (1930), and *Proc. Camb. Phil. Soc.* **22**, 301 (1930).
8 P. A. M. Dirac, *Proc. Camb. Phil. Soc.* **30**, 150 (1934).
9 P. A. M. Dirac, *Nature* **139**, 323 (1937), and *Proc. Roy. Soc.* **165A**, 199 (1938).
10 P. A. M. Dirac, *Proc. Roy. Soc.* **133A**, 60 (1931), and *Phys. Rev.* **74**, 817 (1948).

6

Dirac's way

Rudolf Peierls
Oxford, England

Friends and acquaintances of Paul Dirac were often struck by his surprising and sometimes 'odd' reactions on topics arising in a conversation. Yet, when you had time to reflect, it became clear that his remark was the natural and logical response, and that it was only the automatic and unthinking associations of everybody else which made us expect something different. The same quality can be seen in his physics. The similarity is so close that, as I shall show, many of the famous anecdotes (some perhaps apocryphal, but nonetheless characteristic) can be put in parallel with some of his papers.

Take, for example, the well-known story, told to me by H. R. Hulme, of the pills in the bottle. Hulme apologised for the rattle in his pocket by explaining that a bottle of pills was no longer full, and therefore made a noise. Dirac's comment: 'I suppose it makes the maximum noise when it is half full?' He had seized on the fact that the bottle was silent not only when empty, as is obvious, but also when completely filled. This thought is similar to the idea underlying his 'hole theory'. When I first heard this story, without the date, I thought it would have been very interesting if this conversation had preceded the hole theory, so that the phenomenon of the bottle might have led to the hole theory. However, it was much later, so Dirac merely repeated the train of thought that had given rise to the hole theory.

On another occasion, tea-time conversation concerned the fact that,

43

of the children recently born to physicists in Cambridge, a surprising proportion had been girls. When someone airily remarked, 'It must be something in the air!' Dirac added, after a pause, 'or perhaps in the water'. He had taken the phrase 'in the air' not in its loose conventional meaning, but literally, seeing a possible application. This trend is reflected in much of his work, perhaps first in his picking up Heisenberg's observation that the quantum variables did not commute which, to Heisenberg, seemed an ugly feature of the formalism. Dirac showed instead that it had a very significant place in the new theory.

The importance of looking at the real meaning of a conventional remark extended even to comments on the weather. Jagdish Mehra reports how a visitor, sitting next to Dirac at college dinner and anxious to start a conversation, said, 'Very windy today'. Dirac got up and went to the door, so that the visitor began to fear he had somehow offended him. Dirac opened the door, looked out, and, resuming his seat at the table, said, 'Yes'.

Another characteristic is illustrated by the story that, during a visit to Copenhagen, it was decided that Pauli was putting on too much weight, and Dirac was asked to watch that Pauli did not eat too much. Pauli entered into the spirit of the game and asked Dirac how many lumps of sugar he was allowed in his coffee. 'I think one is enough for you', said Dirac, adding after a moment: 'I think one is enough for anybody'. After some further reflection: 'I think the lumps are made in such a way that one is enough for anybody.'

Such faith in the orderliness of the world is reflected often in his writings, above all in the remark in the paper pointing out that a magnetic monopole would not contradict the known laws of quantum mechanics: 'One would be surprised if nature had made no use of it.'

The same idea is also reflected in his conviction that the true theory must have mathematical beauty. He says in his 1980 paper, 'Why we believe in the Einstein theory' (*Symmetries in Science*, ed. B. Gruber and R. S. Millman, Plenum Press, 1980), that the real basis for believing in general relativity does not lie in the experimental evidence. 'It is the essential beauty of the theory which, I feel, is the real reason for believing in it'.

It is said that he was once writing a paper at the dictation of Niels Bohr. (It was Bohr's usual habit of dictating a draft to one of his collaborators, while walking around the room.) At one point, Bohr interrupted himself and said, 'Now I do not know how to finish this sentence'. Dirac, so the story goes, put down his pen and said, 'I was taught at school that you should never start a sentence without knowing the end of it.'

He seems to have observed this injunction himself. At least, all his papers reflect very great tidiness of expression. Few drafts seem to have survived, but there were probably very few alterations or corrections. When he talked in a seminar about a recent paper, he explained his work almost verbatim in the same terms as in his published paper, the form he had chosen as the clearest expression.

One anecdote is instructive on how legends form. Tyabji, a retired lawyer from India, who was studying theoretical physics under Dirac, discovered that Dirac was very anxious to meet E. M. Forster, the novelist. He invited both to dinner, and when they were introduced, Dirac asked, so Tyabji told me, 'What happened in the cave?' (referring, of course, to Forster's *Passage to India*), with Forster replying, 'Nothing'. This answer apparently satisfied Dirac, and he asked no more during the rest of the evening. I later heard another version, according to which Forster's reply was, 'I do not know.' I thought the earlier version, directly from the host, was likely to be right. Later, when I met Dirac, I asked if he remembered the occasion. He did remember it very well, and both versions were inaccurate. In fact, he had asked whether there could not have been a third person in the cave, which would have meant that neither of the protagonists had told an untruth, and Forster replied, 'Absolutely not. There was no other person in the cave.'

As usual, Dirac had thought of an interesting possibility which had occurred to nobody else. He did this, of course, in so many of his papers, that there would be no point in selecting specific examples.

These few stories might suffice to illustrate my point that unexpected and endearing turns of his conversation were typical of a mind which gave us so many fundamental contributions to physics.

7

An experimenter's view of
P. A. M. Dirac

A. D. Krisch
University of Michigan

When Behram Kursunoglu first asked me to write a short paper about my interactions with Prof. Dirac, I was very honored, since Dirac was probably the most distinguished scientist that I will ever meet. However, I pointed out that it was very difficult for an experimenter, such as me, to understand much of Dirac's work, and I gave a fairly negative response. Nevertheless, I soon recalled how difficult it is to say no to Behram, and I also recalled that he was responsible for providing many opportunities for me to interact with Dirac, which was indeed a valuable gift. Thus, I agreed to write a short article with some anecdotes and perhaps a few thoughts from an experimenter about Dirac's unique view of the world. Let me stress that my quotes from Dirac are only as accurate as my memory.

I first heard Dirac lecture in the 1960s at one of the Orbis Scientiae Conferences which Kursunoglu organized every year at Coral Gables. I was a very young man, and I am sure that Dirac did not notice me. I recall being very impressed by the precise and careful way in which Dirac spoke.

In 1974, I was involved in organizing, at Argonne National Laboratory, what became the 1st International Symposium on High Energy Spin Physics. It seemed that the high energy physicists, who were just starting to study spin, would benefit by hearing directly about the history of spin in the 1920s. Thus, I fearfully but bravely telephoned

46

Prof. Dirac at Florida State University and asked if he would come and give us a lecture on the history of spin. He silently thought for a minute and then said, simply, 'Yes'. I then said that he would be most welcome to spend the entire week and hear something of what we high energy people were trying to do with spin. After another 20 seconds of silence, he said, 'I think that would be interesting'. I was starting to learn that Dirac was a man of few words and that he responded most positively to concise questions.

Prof. Dirac gave use a beautiful and precise lecture on 'An Historical Perspective of Spin' which was published in the proceedings of the Symposium (ANL/HEP 75-02). One technical point about this lecture impressed me enormously. We agreed to tape-record Dirac's lecture and then to have it typed by a secretary at Argonne. As many of you know, transcribing lectures is normally a very painful process, since few of us speak in sentences. For Dirac's lecture, the process was trivial; Dirac spoke in perfect sentences. If you are not impressed by this, listen sometime to a recording of one of your own lectures.

One day, I asked Dirac if he would like to see the 12 GeV ZGS (Zero Gradient Synchrotron) Accelerator, whose new polarized proton beam was really the reason for the Symposium. He said immediately that he would be very interested in seeing it. After several minutes of silence he said that he '. . . had visited several other accelerator laboratories but had spent essentially the entire visit in the Director's Office, and no one had offered the opportunity to see the accelerator.' We arranged for Charles Potts, the engineer who was in charge of ZGS operations, to show Dirac the ZGS. They spent more than an hour on the tour. Afterwards, Dirac told me that he had a '. . . very interesting time with Mr. Potts.' He then volunteered that his '. . . undergraduate education was in engineering, and I went into theoretical physics because in the 1920s, when I graduated, there was a recession in England, and I was unable to find a position.'

A second interesting incident occurred at the Symposium Banquet, where we had not arranged an after-dinner speaker. About 20 minutes after coffee was served, the conversation was beginning to slacken, and I was getting ready to suggest that we should thank the Argonne Cafeteria for a fine Roast Beef dinner and go to bed. Suddenly, Dirac stood up and said, 'I think that it is an excellent tradition at Conferences to tell amusing stories about physicists, and I would like to tell one if there is no objection.' Of course, no one objected, and he then told us a story, which I think was about Kemmer. Unfortunately, I do not remember the story. From this incident, I came to see that Dirac was not excessively shy, he just did not speak unless he had some point to make.

We have continued Dirac's tradition of telling amusing stories through the 6th High Energy Spin Physics Symposium at Marseille in 1984, and I hope that our Russian colleagues will continue it at the 7th Spin Symposium at Serpukhov in September 1986. I would like to add that many of the amusing stories have been about Wolfgang Pauli. According to C. N. Yang, such stories may be especially appropriate at a Spin Symposium in view of Pauli's lack of enthusiasm for spin when it was first proposed. This lack of enthusiasm was certainly well documented by Dirac in his Argonne lecture.

The third incident involves what may be the most famous Dirac story.

Dirac gave a Colloquium somewhere in England. At the end of the Colloquium, the Chairman asked if there were any questions. Prof. X then stood up and said, '...I do not understand the last equation which Prof. Dirac wrote'. There followed several minutes of silence which became increasingly uncomfortable. The Chairman then asked if Prof. Dirac would care to answer Prof. X's question. Dirac replied, 'He did not ask a question, he made a statement'.

On the day after the Symposium Banquet, while walking through the Argonne woods on the way to lunch, I related this version of the story to Dirac and asked if there was any element of truth in it. He walked on in silence for about a hundred yards with his hands clasped behind his back. He then said, 'Perhaps there was some element of truth in it'. After another hundred yards of silence he announced, 'After all, it was a question of professional competence; the man claimed to be a theorist.' From this incident, I learned that Dirac also had high standards for others.

In 1978, we decided to have a series of occasional public lectures at the University of Michigan by very distinguished physicists. Recalling Dirac's Argonne lecture, I thought that his giving the 1st H. R. Crane Lecture would certainly set an adequately high standard for the lectures. I again found that by inviting him directly, I got the same simple 'Yes' after a brief silence. However, after some thought, he expressed concern about the possibility of snow in Ann Arbor, so we arranged the Lecture for April 17, 1978, which is near the end of our semester when the probability of snow is small. He also asked if he could bring Mrs. Dirac, since he '... was finding traveling alone more difficult in recent years.' We, of course, were very pleased to have them both, and, during their visit, we came to appreciate Mrs. Dirac's unusual intelligence and understanding of people. She is, of course, one of the rare people in the world whose brother and husband both received Nobel Prizes.

Dirac's lecture on 'The Prediction of Antimatter' was superb. It was attended by about 1500 people, many of them students who came from throughout the state of Michigan. His lecture was, of course, concise and logical. It also gave an excellent picture of what I consider perhaps the greatest triumph of pure theoretical physics. I am unaware of any hint of the existence of antimatter prior to Dirac's prediction. The elegant and simple Dirac Equation suggested the existence of antimatter, and it was found as predicted.

An amusing anecdote occurred during this visit. One morning, Dirac stated that, when he had '... last visited Michigan 49 years earlier (1929), Prof. Randall had a very high quality ruling machine for making diffractions gratings, and would like to see it again.' When I indicated that I did not know where it was, he said that he remembered and would show me. He then led the way to the 2nd sub-basement of Randall Lab., where he, indeed, found the special vibration-free room. Unfortunately, the ruling machine was no longer there, and we only found an undergraduate asleep on a lab table. The student's expression when he awoke and found Dirac inspecting him was interesting.

For almost a decade, I attended most of the Orbis Scientiae Conferences in Coral Gables, and I believe that Dirac attended all of them. I had many opportunities to interact with the Diracs, because Behram Kursunoglu often arranged invitations to small dinner parties held by his prominent friends in Miami. Of course, Dirac was the guest of honor at all of these parties. One interesting incident occurred in a reception line at one of these parties. One of the guests reached Dirac and said, 'Oh, Professor Dirac, what does it feel like to be the most brilliant man in the World?' Dirac stared at her with his penetrating eyes for several minutes and was silent. As the minutes passed, she started giggling more and more nervously and then quickly shook my hand and more quickly left the room. Dirac then turned and said, 'What

49

can you possibly say to someone like that?' I decided not to answer. I learned something from this incident which I have found useful in Committee meetings.

Another anecdote involves a lecture that Dirac gave at the 1977 Orbis Scientiae. Behram was being even nicer to me than usual and appointed me as Chairman (Moderator) for Dirac's lecture. Behram had outdone himself in getting distinguished participants at this particular Orbis, and Richard Feynman was sitting in the front row. Dirac decided to devote much of his lecture to pointing out the various shortcomings of QED, the theory for which Feynman had received the Nobel Prize. Dirac emphasized his belief that a theory with such inelegant infinities, which had to be subtracted, could not possibly be correct. As Moderator, I was looking forward to the question period with some concern, since Feynman is not noted for his mildness, and I felt somewhat out of my depth in this company. However, Feynman merely nodded quietly each time Dirac referred to these inelegant subtractions, perhaps with some disdain. What I learned from this incident was that either Feynman shared Dirac's concerns or that there may be levels in the Theoretical 'pecking order' that are not easily observable to an experimenter.

Behram Kursunoglu encouraged me repeatedly to say something about my interactions with Dirac related to our experiments on spin effects in very violent proton–proton collisions. Unfortunately, I cannot recall ever discussing the experimental data with Dirac. This may be because of the many times that I saw Dirac approached by eager physicists who enthusiastically told him of their latest important work; a 'veil of weariness' often seemed to descend upon Dirac. I assume that this was because, during the previous six decades, several hundred other 'young' physicists had tried to impress the 'great man' with the fundamental importance of their work. On the day following an Orbis talk, which I gave on the ZGS Polarized Beam, Dirac did ask some detailed questions about the accelerator physics and the hardware used in jumping through the depolarizing resonances; we discussed such things several times in the following years. In his 1974 Argonne paper, Dirac also discussed briefly the similarity between the ZGS polarized beam and the electrons in an atom with respect to the Thomas precessional frequency. I think that Dirac continued to have a lively interest in science and technology; he had just become understandably jaded with people trying to impress him.

An experimenter's view of P. A. M. Dirac

To partially respond to Behram's request, I will say that I certainly find our high energy spin experiments quite fascinating because of their exploratory nature and unexpected results. This fascination comes partly from the long held belief of most high energy theorists that spin would become less important with increasing energy and transverse momentum; this belief now appears to be quite false. Using recent advances in both Polarized Proton Beams and Polarized Proton Targets, we have been able to study spin effects in more and more violent proton–proton collisions, first at the 12 GeV Argonne ZGS and, more recently, at the 28 GeV Brookhaven AGS. These large transverse momentum elastic scattering experiments indicate that both spin–spin and spin–orbit forces appear to be quite large and possibly growing. Such spin effects were certainly not predicted and are, indeed, quite difficult to reconcile with the now-popular theory of Quantum Chromodynamics, which pictures each spin-$\frac{1}{2}$ proton as being constructed of three spin-$\frac{1}{2}$ quarks.

My approach to such an unexpected situation is to explore it further by varying all possible parameters and observing what happens. Thus, we are now looking at even more violent collisions by increasing the AGS beam intensity, which may let us increase the transverse momentum squared to 7 or 8 $(GeV/c)^2$. We may also increase the incident energy to 800 or 900 GeV by doing a similar experiment at Fermilab. This is clearly an inductive or 'exploratory' approach to studying the laws of nature, which may be a proper approach for an experimenter. Our common interest in spin was probably responsible for much of my interaction with Dirac; however, my approach to science is certainly quite different from his.

I recall one scientific discussion with Dirac in Coral Gables on what is the proper technique for searching for truth in science. Dirac stated that, '. . . the elegance of the formulation was very important in choosing the direction for one's research.' I replied that the Scientific Method suggests that the only absolute criterion for truth is the reproducibility of experimental observations. I suggested that this one clear criterion for truth has resulted in enormous progress in science in the past 2000 years, while other fields, such as philosophy and literature, which used elegance as a major criterion, have made relatively little progress. I think, from his slight nod, that this made some impression on Dirac, but I am not sure, since there was only silence for the next 15 minutes. In thinking about this discussion later, I have come to believe that the reproducibility of experimental observations may, indeed, be our only absolute criterion of truth, but the creative elegance of rare people such as Dirac may sometimes point us to that truth. I remain overwhelmed by his prediction of antimatter.

In another discussion, Dirac suggested the possibility that little further progress will be made in theoretical physics until someone invents a new type of mathematics that is capable of dealing with the next level of understanding our Universe. He seemed to believe that our mathematical abilities often limit our capabilities to discover and understand new natural phenomena. In thinking about the difficulty of understanding the hydrogen atom without our knowledge of calculus, I guess that I see his point. I sometimes think that there may be a simple and elegant way to understand strong interactions, including their unexpected spin effects, but that we are blind to it because of our mathematical weakness.

In summary, I found it an unusual benefit to have these opportunities to interact with Dirac. There were certainly many sides to Dirac and his approach to Theoretical Physics that I will never understand and perhaps few others will. However, I did learn three important lessons from Dirac:

1 Physics should be elegant. If the equations are not simple and elegant, they are probably wrong.
2 Never say that anything is true, unless you are certain that it is true.
3 There are talkative people in the world who sometimes violate lesson 2 and do not understand lesson 1.

Lesson 2 is especially important to an experimenter, as it is remarkably easy to make totally or partially invalid assumptions. I will, therefore, close with one of our Spin Symposia stories about Dirac and Pauli which is probably fictitious, but so beautifully illustrates Lesson 2 that I will still tell it.

Pauli and Dirac were riding on a train in the countryside. After an hour of silence, Pauli was becoming uncomfortable and was eagerly searching for an opening remark. Upon passing some sheep, he said to Dirac, 'It looks like the sheep have been freshly shorn.' After studying the sheep, Dirac replied, '. . . at least on this side.'

8

Dirac at the University of Miami

Henry King Stanford
University of Georgia

I remember vividly the first time I met Paul Dirac. He had come to the University of Miami to accept the first J. Robert Oppenheimer Memorial Prize, initiated by the Center for Theoretical Studies. I was not quite prepared for the fragile frame that encompassed such a giant intellect. This shy, wispy gentleman sat at the dinners or stood at the receptions given in his honor, scarcely volunteering a comment until spoken to first. The ordinary chitchat of the social situations seemed painful to him. He would pause before responding to a question as if weighing the consequences of his answer. Always, he seemed preoccupied, once to the point of falling asleep at the dinner table, as if to prove that the conversation he was trying to endure was entirely soporific.

When Paul Dirac stood up to speak, however, he became a different person. Here, he was in the arena of ideas where he could tilt with the best minds and keep them on the defensive. The first speech I heard him give followed his acceptance of the Oppenheimer Prize on March 12, 1969. It was largely autobiographical. The speech was memorable for me, not a physicist, because of the emphasis Dirac placed upon the human emotions attendant upon scientific discoveries. He was intrigued with the role played by 'hopes' and 'fears' in the establishment of new principles. Young scientists, so Dirac conjectured in his talk, would postulate a new theory with high hopes of its acceptance as a new

53

paradigm. Then, sometimes, they would fail to take the 'last, and rather small, step' to assure the logical outcome of fear that the theory would collapse if extended. Dirac cited Heisenberg's idea concerning the construction of a theory in terms of quantities provided by experiments rather than building it up from an atomic model.

Dirac suggested that Heisenberg was fearful of taking the additional step leading to the idea of noncommutation because his theory might collapse. Dirac, however, explained that he, himself, not being the author of the new theory, had no personal stake in it and, therefore, was able to extend it, because he was 'not afraid of Heisenberg's theory collapsing.' In other words, Dirac seemed to be saying that an investigator's or thinker's stake in the 'intellectual establishment' might make him reluctant to accept the 'new' because of his fear of losing the 'old,' particularly if he had had a share in postulating the 'old.' Another example of this hope/fear tension suggested by Dirac was the case of Brackett. In spite of the fact that he was the first to obtain hard evidence for the existence of a positron, he was 'afraid to publish it,' thus, leaving to Anderson the opportunity to scoop him.

The fear that something will go wrong is not confined to physics alone. Fear of the unknown has made human beings cautious since the emergence of our species. The 'establishment,' whether in science, politics, business, or what not, does not usually like boat rockers with untested and unproven notions.

Paul Dirac began his four-year residence at the campus of the University of Miami just as the national student 'boat rocking' appeared at our university. For several years, students on campuses across the nation had demonstrated in favor of, or protested against, a number of issues: 'free speech,' beginning at Berkeley; the war in Vietnam; civil rights, including greater opportunities for women and minority groups; consumerism; and so forth. Somehow Dirac seemed drawn to the public demonstrations and even sit-ins which characterized that era on the campus of the University of Miami and other universities across the United States. He was never one of the demonstrators, to be sure, but I observed him at several demonstrations, standing on the periphery of the protesting groups, keenly listening to the student charges and proposals and my response to them. The dialogue between the students and me on 'The Rock,' a stage-like stone structure in the middle of the campus, fascinated him.

Dirac at the University of Miami

In May, 1970, I was hurrying to 'The Rock' to speak before 7,500 students who had rallied to protest the tragedy at Kent State University in Ohio. There, several students had died two days before from the shots of Ohio National Guardsmen. Professor Dirac suddenly stepped out from the edge of the crowd, approached me, and asked: 'Are you afraid?' When I replied that I was not and added, feeling the adrenalin flowing, that I was looking forward to the exchange, he then offered this advice: 'Tell them what you think and listen to what they have to say.' That was what I was going to do anyway, but it was surprising and reassuring to receive this practical advice from such a renowned scientist.

At the time, I thought I detected a spiritual kinship in Dirac's mind between himself and the students. I do not mean that he agreed with all the issues they were raising with me. Rather, in advising me to engage in a sincere exchange with the students by listening to what they had to say, he was supporting implicitly the right to challenge 'the establishment,' of which I was the foremost symbol on the campus.

I had the firm conviction that he was putting himself back into the youth of his early twenties when he was challenging another kind of establishment, the accepted physical theory of his day. I remembered his Oppenheimer lecture. The theoretical and experimental workers of the 1920s, the physical establishment of his youth, 'would look for any explanation rather than postulate a new particle,' he reminisced in the lecture. But Dirac enjoyed something then that I used to tell my students they enjoyed, namely, the 'advantage of inexperience.' Advancing years have a way not only of piling up disappointments, disillusionments, and even failures, but of freezing a mind-set as well, of turning 'what is' into the 'ultimate.' Among my students, idealism and enthusiasm abounded. The white charger was always present, ready to be mounted, ready for the charge when new worlds would be conquered, never windmills tilted. The students did not know what would not work. 'Don't tell us it can't be done,' they seemed to be saying; 'we have not tried to do it.'

I am sure that that way of thinking, the oblique think in comparison with the direct thought of traditional processes, was the road that led young Paul Dirac to challenge the accepted notions of physical theory of the 1920s. In that first lecture on our campus, he claimed two advantages over Heisenberg: lack of fear of upsetting Heisenberg's notion because he had no vested interest in it and the youth of a research student. Dirac's mind was not immersed in the ultimate reality as the

55

contemporary intellectual establishment perceived it; so he was free to mount his own charger and ride off to conquer new worlds of physical theory. His keen interest in the student issues and confrontations of the late Sixties in the United States was stimulated by nostalgia, I believe, for the days when he, himself, was postulating ideas that had not occurred to the intellectual establishment. He wanted me to listen to the young as he, himself, had been listened to.

9

Remembering Paul Dirac

Eugene P. Wigner
Princeton University

Scientific accomplishments

I want to tell, principally, about the early years' accomplishments of
Paul Dirac – he was less famous in those years, and you may know less
about them. The picture I'll present may also convey his very modest
and considerate personality. As you probably know, he was born in
England – in Bristol in 1902, just a few months before I was born in
Hungary. He had a brother, older than himself, and a sister, younger
than he was. But he hardly ever mentioned his family – he believed
others were not truly interested in them. He did learn French from his
father, who was a teacher of that language, but he did not like to use it.
He had profound admiration for few people, not even for his father.

But he did like the Bristol high school he attended, called Merchant
Venturer's School. It seems that the teachers had appreciation for the
gifted students and helped them individually by providing support for
their interests. Perhaps I may mention that this was true also in the high
school I attended – our mathematics teacher gave private classes to John
Von Neumann and gave me interesting mathematical books to read.
Actually, the Merchant Venturer's School was closely connected with
the University of Bristol, and it was natural for Paul to continue his
studies there after graduation from high school.

What was, perhaps, less natural was that, in spite of his early and

57

overwhelming – and eventually marvelous – interest in mathematics and physics, he studied electrical engineering. The reason may have been, perhaps, similar to my own reason for studying chemical engineering – the chances were small of obtaining a job as a physicist. He did not really confirm this reason in the course of our conversations. But, as he emphasized, his study of electrical engineering had some good consequences (as also did my knowledge of chemistry): he became aware of the fact that the laws of nature which man can discover will be approximate laws – just as the designs of the electrical engineer are based on approximate rules. He felt that recognising the usefulness of approximations – and the impossibility of the discovery or use of totally accurate laws of nature – had a fundamental and very useful effect on his thinking. He told this to me in the course of our conversations many years later – when we had close contact at Princeton University.

He admitted, however, that, in spite of his prime subject of electrical engineering, he also studied physics and had the moral, though rarely direct, support of his teachers in this regard. He graduated in 1921 but remained in Bristol for another year, learning mostly mathematics and physics. As to the latter subject, he concentrated on general relativity and studied quantum theory only superficially. In mathematics, he was interested in approximation methods though, to be frank, I never understood in which way.

I should admit here that most of the preceding information about Paul Dirac's early life was furnished by my recollections of early conversations with him – I hope my recollections are correct, but I'll tell more about the occasions when we had those conversations.

I feel I should now review briefly the origin of quantum mechanics, because the original work introducing totally new ideas had an enormous and lasting effect on Paul.

Before the so-called quantum mechanics was created, there was, in the minds of most physicists, some question of whether man is bright enough to overcome the internal problems of quantum theory – whether the problems of the microscopic structure of matter and of radiation are simple enough to be solvable by the human mind. This question was not publicly discussed, but, attending the physics colloquia at the University of Berlin, I had the impression that most of the famous participants, including Einstein, Planck, Von Laue, and several other great minds, had such doubts – just as I am unsure now whether the existence of life, of human knowledge and emotions, can be incorporated into a fundamental extension of the present theory of nature. The quantum theory of matter was, at the time referred to (before 1925), in serious trouble. It could not describe any atomic structure except that of hydrogen, and even the description of hydrogen was clearly inadequate. These were the problems for the solution of which Heisenberg produced (in 1925) the key: he proposed that Bohr's picture of atoms, though admirable, should be replaced by a theory based on observable quantities: the energy levels and the transition probabilities between these. For hydrogen, these were given quite well by Bohr's picture of electrons, whose state was classically described by the paths they traveled around the nucleus. But this classical picture could not be extended to other atoms, and Heisenberg proposed (July 1925) the radical theory to replace the classical picture by one which used only observable quantities: energy levels and transition probabilities.

Heisenberg's wonderful idea and his proposed theory were enthusiastically accepted and formulated in detail by Born and Jordan – very soon (just about a month) after Heisenberg proposed his theory. Dirac, I believe independently, accepted Heisenberg's proposal to abandon the picture of electron orbits, in fact, the classical description of the state of the electrons, and proposed, in December 1925, a quantum mechanics which was inherently very similar to that of Born and Jordan and also to the later (February 1926) article of Born, Heisenberg, and Jordan. He also extended his December article by one in March 1926 which could already be compared with the Born, Heisenberg, and Jordan article.

These were wonderful developments which convinced, I believe, all physicists that, after all, man is bright enough to create a physics of the microscopic, that is atomic, phenomena. Of course, they were still far from that – they gave, essentially, only ways to calculate energy levels and the probabilities of transitions between these caused by the absorption or emission of light – but they were wonderful to do that much and they were relatively simple. I must admit that Dirac's papers, though highly respected, had less influence than those of Heisenberg, Born, and Jordan. It was not easy to fully digest them, and this diverted some attention from Dirac's work which was also quite difficult to understand fully – partly because it was in English, while most physicists of interest in the subject were Germans.

But, I believe, every physicist interested in quantum theory knew about his work and admired it – even though it was, perhaps, less easily digestable than those of the three German authors. And, in some regards, it went further than those of the other writers.

The next fundamental step forward cannot be attributed either to the German trio (Heisenberg, Born, and Jordan), or to Paul. It came, at least for most of us, unexpectedly, from one who had not contributed fundamentally to quantum theory before. It came from Erwin Schrödinger – only five years younger than Max Born, who often doubted the possibility of making basic scientific contributions at his age. Schrödinger, in March, April, and May 1926 articles, reformulated quantum mechanics, giving a new description of the states of the physical systems. According to the earlier description of physical reality, the energy of the atom is always definite and corresponds to one of the energy levels. Schrödinger's description of the 'state' of a system includes linear superpositions of such states, it postulates an infinitely greater variety of states than did the ideas of the earlier contributors. Perhaps I should formulate this more explicitly, though probably all of you know it: Schrödinger characterised the state of the system by a 'wave function'. This can be written, if so desired, as a linear combination of the wave functions which correspond to states with definite energy values, the coefficients being complex numbers. The absolute squares of these complex numbers do give the probabilities that the measurement of the energy gives the energy value which corresponds to the wave function of which the number is the coefficient. But not only the absolute value, but also the complex phase of the coefficient is relevant – the outcomes of other types of measurements are decisively influenced by the complex phases of these coefficients. This means that the multitude of the possible states of the system is, already according to Schrödinger's theory, much greater than it is in classical physics, whereas, it was much smaller in early quantum mechanics which recognised only states with definite energies – or possibly mixtures of these. Actually, Schrödinger's wave functions could describe not only energy values and transition probabilities – as was postulated by Heisenberg for the initial quantum mechanics – but a much more general situation and behavior which could approach classical mechanics at higher energies. It also led to the possibility of describing internal changing situations, not only those which consist of the absorption or emission of light. It describes also chemical and nuclear reactions which play, of course, a very important role in our physics. I would like to mention also that their description of collision processes – which can lead to chemical or nuclear reactions – led John A. Wheeler to defining the so-called collision matrix, which is, according to the late ideas of Heisenberg, the most important observable quantity of microscopic physics. In summary, I'd like to state that Schrödinger's 'second equation', giving the time dependence of the wave function, and, hence, of the state of the physical system, was (and is) of most fundamental importance. It was so recognised quite soon, in particular also by Dirac.

The development of quantum theory through the contributions of the aforementioned five contributors (including Dirac) was fantastic. Naturally, in addition to the aforementioned five, there were other highly effective contributors (including W. Pauli). But the most important contributions in the next two years, 1927–28, can be attributed, I believe, to Paul Dirac. The first of these is the introduction of quantum field theory (February–March 1927).

As already implied, the interaction of light with matter was not described adequately by either of the theories proposed by 1927. In fact, no theory was available which would deal with the quantised nature of light and its interaction with matter as a result of its quantum nature. All this was provided by an article from Paul Dirac in February and March 1927. He introduced the idea that some components of the electromagnetic field strength are observable at definite space–time points of a space-like surface and quantised the resulting field in such a way that, if its local periodicity corresponds to a definite wave-length, its total energy, determining the time dependence of its state vector, corresponds to an integer multiple of the value given by Planck. The expression for the Hamiltonian interaction with charged particles is the classical one. It would be too complicated and lengthy to give explicit details of his article (*Proc. Roy. Soc.* **A114**, 243 (1927)), but it is well known that it was very, very successful. It was successful not only in describing the light-absorption and emission processes but was followed also by the description of other somewhat similar processes, including that of the β-decay, and several others contributed by physicists particularly interested in them and in the 'field theory' founded by him.

It should be admitted in this connection, nevertheless, that these field theories do somewhat lack beauty and simplicity. They are usually applied by expanding the mathematical expression for the rate of processes – in our case, the emission and absorption of light quanta – and the effects of these processes on the interaction between the particles which emit and absorb them. But the result of such calculations often gives infinite – and, hence, incorrect – results. The infinities are then eliminated by the process called renormalisation, which is necessary and useful, but greatly decreases the mathematical beauty of the theory. It must be admitted, on the other hand, that, up to the present, no idea with usefulness similar to those of the field theories has been invented. We hope it will be and the postulate of relativistic invariance will make it easier, not more difficult, to invent it.

Dirac's next basic contribution came in 1928 and was of a very different nature. He proposed a new equation, again a relativistically invariant one, but for a single particle, the electron. I will not describe this equation, as all physicists are familiar with it. But I will tell you that just about the same time that Dirac's electron equation was discovered by him, Pascual Jordan and I also tried to find the correct relativistically invariant equation for the electron. We worked hard on the problem and were considering several equations but, frankly, we did not truly like any of them. Then one day, Born, the professor of theoretical physics at Göttingen, received a letter from Dirac, relating to his proposed visit to Göttingen. And there was a postscript to the letter which told, in about four lines, about *his* electron equation. Born showed the letter to Jordan and he was overwhelmed. Dirac's equation differed from all those which we were considering by having four components, whereas, all those we considered had only two, corresponding to the two possible components of the spin in any direction. But Jordan was, at once, overwhelmed by Dirac's equation, and, when he told me about it, he said, 'It is too bad we did not discover that equation, but it is wonderful that someone did.'

As you probably also know, Dirac was perturbed by the fact that his equation had solutions with negative energy values. He assumed that the states of negative energy are occupied, so that no particle can be moved into a negative energy state. If a negative energy particle is moved into a positive energy state – the hole so formed will have a positive electric charge – some negative charge will be missing. He first thought the resulting apparent particle was the proton, but, when Carl D. Anderson, in 1932, discovered the positron, he realised that the apparent hole is a positron, a particle with the same mass as that of the electron. Dirac's electron equation is not only mathematically neat, its various consequences, in particular, its responses to potentials such as that of a proton, have also been confirmed.

My discussion of Dirac's later contributions to physics will be very short and superficial. Actually, in most cases, a satisfactory review of an article would take as much space as did the article itself. But I would like to mention the article which impressed me most. It deals with the large numbers in physical theories and in the structure of the world. The ratio of the electromagnetic and gravitational forces between a proton and an electron is about 2.2×10^{38} – a fantastic number, but the number of protons in the universe and the alleged age thereof in terms of h/mc^2 are even more fantastic. Actually, he suggests that these numbers are related, but he defines two time measures for the distant past, and the age of the universe, the time when the big bang occurred, is infinitely long in terms of one of them.

I consider the article just referred to to be exceptionally interesting – even though it does not share the characteristics of other surmises in physics which can be tested experimentally. But he wrote several other semi-philosophical and very interesting articles which I will not – and, mostly, cannot – review.

This concludes my review of the scientific contributions of Paul Dirac – it is, naturally, a very superficial review. But a detailed one would not be really possible. I will now go on to talk about his personality and his relationships with his colleagues, principally myself – I am more familiar with that than with his friendship and collaboration with other colleagues.

'My famous brother-in-law' – our personal relationship

I first met Paul in 1928 in Göttingen – he was invited to give an address on his recent work, I believe on his original ideas which led to quantum field theory. He was, scientifically, remarkably clear and detached from any personal remarks – almost as if he had read to us from a publication or a book. This remained with him throughout his life – he seemed to be more interested in science, and the particular problem he was working on, than in himself. But we enjoyed his presentation, even though the discussion which followed it was not very thorough. His answers to questions, as on later similar occasions, were always restricted to the problem raised by the question – he did not extend his answer to subjects related to the question raised. This restriction remained with him almost completely for all later years. I did mention an exception: the four added lines to his letter to M. Born, announcing his relativistic theory of the electron, were not related to his visit planned at that time.

Our relations in later years were much closer and much more extended. During the period of my half-year visits to Princeton (1931–6), he often spent a few weeks there; after 1932, at the Institute for Advanced Studies. We had our meals together, almost constantly, at one or another restaurant, again with rather little personal conversation. It was, nevertheless, on these occasions that I learned about his family, his rather weak attachment to its members, his past interest in mathematics and even in electrical engineering, and about himself. He seemed to like some company, even though silent company seemed to be his preference. But, during our many, many joint meals, I started to know about his past and to like him not only as a big contributor to science, but also as a friend.

I like to recall two particular incidents of our joint meals. Once, a third colleague joined us, and there was a problem on which we had different opinions. We both argued vigorously but, well before we came to an agreement, he had to leave. Paul seemed to listen in but did not voice his opinion, not on a single occasion. When we were left alone, I asked him why he did not speak up? His answer expressed his conviction about the usefulness of arguments, such as those which the two of us put forward. He said: 'There are always more people willing to speak than willing to listen.' And he had adopted that useful role on this and many other occasions.

The second story is very personal. In 1934, my sister, subsequently Mrs. Dirac, accompanied me on my Princeton half-year sojourn. Her marriage had broken up – her husband abandoned her, and she was very unhappy. I thought that a new environment might divert her from her painful experience.

One day, when we visited an eating place, she was surprised by our having, apparently, attracted the attention of a new visitor to the restaurant. I did not see him because I was facing the opposite direction. But she asked me to look back. I did and saw the unexpected visitor, Paul Dirac. I told my sister that we used to have our meals together when both of us were in Princeton. She then asked me to invite him to join us, and he did. Paul and my sister became acquainted this way, and they married, to my great pleasure and some surprise, on January 2, 1937. He became a very loving husband and adopted my sister's two children from her earlier marriage. They also had two children.

Of course, that marriage greatly strengthened my interest in, and my affection for, Paul Dirac, and we met quite often after that. As I said, he was a very affectionate husband, and their two daughters were fond of him and are still very fond of my sister. I appreciate the lucky life which my sister gained from her marriage to Paul, and his departure from our world sorrows me greatly.

2

MORE SCIENTIFIC IDEAS

10

Another side to Paul Dirac

R. H. Dalitz
University of Oxford

1 Introduction

My first contact with Professor Paul Dirac came in January 1947 when I began to attend his lectures on Quantum Mechanics in the Arts Faculty Building, Cambridge University, as a research student in mathematical physics. I had already read the first edition of his book *The Principles of Quantum Mechanics* during our six-week voyage by refrigeration ship from Sydney to London in 1946. In Cambridge I soon learned that much of the book had been rewritten and that the third edition was then in course of publication. Dirac lectured to us by reading from the proofs of the new edition and kept us very busy as we noted down the changes and the new text. This edition introduced the bra and ket notation which pleased Dirac so much.

The class attending Dirac's lectures was rather large. Not only were there many ex-servicemen, returning war workers and students from abroad who had had to delay their coming to Cambridge, but also interest in quantum mechanics had grown rapidly during the war years because of its widespread applications. S. Shanmugadhasan once mentioned to me that, when Dirac came to his first lecture in 1945/6, he was so astonished at the crowd waiting in the lecture room that he said to them 'This is a lecture on *quantum mechanics*', supposing that most of the students were there by mistake and would then leave. Since nobody

69

left, he repeated this statement more loudly, but still nobody moved, so he had to begin his lecture to this unexpectedly large audience. In 1946/7 there were also many students attending Dirac's lectures for the second time in order to hear the text of the new edition. However, it was quite common for students to attend the course twice, even when the text was the same.

2 Dirac's engineering training

Most people are aware that Dirac had been an engineer before he came into theoretical physics and became quickly prominent as one of the creators of quantum mechanics, although the stories given in the literature are often not correct in detail. These stories interested me particularly, since I also began my training as an engineer. In a developing country like Australia, the engineer was clearly an important person and what he achieved was visible to all. Nobody was able to say where a training in mathematics and physics would lead, other than to teaching; the pure scientist was invisible and unknown, and entry into those specialities was discouraged both by University advisors and by parents. So I played safe and entered the engineering track. After I heard the stories about Dirac, I wondered whether he had been impelled by similar forces.

The Merchant Venturers' Technical College at Bristol had been established as such in 1885 by the Society of Merchant Venturers of the City of Bristol, their aim being 'to provide a sound, continuous and complete preparation for an Industrial Career'. It had both Senior and Junior Classes, the latter constituting a four-year Secondary School for Boys. Dirac's father Charles was the head of French language teaching in the College when Paul entered the Secondary School in the Autumn term of 1914, at age 12 years. Charles Dirac lectured in all parts of the College, including the additional evening classes. In 1909, the University College of Bristol gained its charter to become the University of Bristol and the new University decided to enlarge its Faculty of Engineering by incorporating the Senior Engineering Classes of the Technical College; the arrangement was that the Principal of the College became the Dean of the Faculty of Engineering in the University. Since Charles Dirac regularly taught French to the Senior Classes, he thereby became a Recognized Teacher of Bristol University in the new Faculty.

It was natural for Paul to attend this school, as had also his brother Reginald, older by two years. It was quite a large school, giving a thorough training with a practical bent. Modern Languages were taught for use, there was some History and Geography, but not Classical Studies. This gives point to a story of a later time, at St. John's College, Cambridge, when Cockcroft (who had also trained as an engineer), after a discussion at the head of the dinner table, called to Dirac in the middle of the soup course: 'Do you consider yourself to be an educated man, Dirac?' Dirac did not reply, but kept on eating. At the end of the third course, he replied to Cockcroft in a low voice: 'No, I don't know any Latin or Greek'. However, Paul certainly stood out as a mathematician in the Secondary School and spent much of his time in more advanced studies, separate from the rest of the class. The School was fortunate in being part of the College, the well-equipped Senior laboratories being available for use by the School, being especially accessible in those War years when the senior students were rather few.

It was also natural for Paul to stay on in the same buildings and to study engineering. His father recommended this course, as he had done for Reginald, who had wished to study medicine, and there was no one to object. Reginald had already been studying mechanical engineering for two years, and so Paul followed after him, specializing in electrical engineering and graduating with first class honours in 1921. Despite this success, he was not considered a promising engineer, because the report from the British Thompson Houston Works at Rugby where he had worked to gain the practical experience which was required before graduation, was not favourable, as he himself recalled in an interview[1] in 1963 and as his contemporaries (several of them still alive today) also remember. As a result, he did not find a job in engineering and it is clear that such work did not really hold his interest. To fill in time, David Robertson, the Electrical Engineering Professor at Bristol University, encouraged him to start work on some practical projects in his department. At this point, the Mathematics department staff, who had been rather unhappy when Paul chose to go into engineering rather than mathematics after he left the Secondary School, proposed that he should follow the Mathematics lectures, unofficially and without tuition fees, covering the course in two years. Since he was able to live at home, this was feasible financially and attractive to him personally, and so he accepted this offer.

Dirac's own recollections from that period stress the great attraction Projective Geometry had for him, in large measure because of its beauty and the great extent of the implications of rather few assumptions, but also because of the skill of the teacher, Mr. Peter Fraser. There was only one other student following the honours course, Miss Beryl May Dent, daughter of a schoolmaster at Warminster, near Chippenham, who was a registered fee-paying student. There was a choice of specialization in the final year; as the only official student, Miss Dent had the right of choice and she decided to opt for Applied Mathematics. Since the Department could offer only one set of lectures, it was therefore necessary for Paul to follow the Applied Mathematics sequence also. At the final examination in 1923, Paul and Miss Dent both gained first-class degrees and they were both accepted as research students by Cambridge University, Paul at St. John's College and Miss Dent at Newnham College. Paul had hoped to be taken on by E. Cunningham, to work on relativistic electrodynamics, but this was not possible at the time and he became a student of R. H. Fowler instead.

3 Dirac's practical side

Besides all of the major research which he did on the formulation and development of quantum mechanics, on the incorporation of relativity into quantum mechanics leading to his greatest achievement, the Dirac equation for the electron, and on the founding of quantum electrodynamics, Dirac did make mundane calculations of cross-sections or rates needed by experimenters, from time to time. For example, he was the first to calculate the cross-section for the annihilation of a positron by a stationary electron giving two photons. This paper[2] actually bears the title 'On the annihilation of electrons and protons', since the positron had not yet been identified empirically at that time and Dirac identified the electron hole with the proton. However, Dirac carried the calculation through for a single mass value m, commenting only that 'we do not know whether the m there refers to the electron or the proton. Presumably it is some kind of mean.' When m is taken to be the mass of the electron, the cross-section formula is that now quoted for the process $e^+ e^- \to \gamma\gamma$ in the standard works on electromagnetic radiation processes[3]. In 1932, he published some calculations of photo-electric absorption cross-sections[4] carried out jointly with J. W. Harding, who was then a student of Rutherford, work prompted by the questions of the experimenters engaged in measuring these processes in the Cavendish Laboratory at Cambridge[5]. In 1933, jointly with P. Kapitza, he discussed the characteristics of a very interesting physical situation[6], the diffraction of electrons from a 'grating' consisting of standing light waves, a process whose observation they showed to be outside the range of experimental possibility at that time. It is of interest to recall here that Kapitza and Dirac gave a review of this proposal and of the possibility of carrying out the necessary observations to check it, at an extraordinary meeting of the Kapitza Club on 10 May 1966 during Kapitza's visit to England to give a lecture about Rutherford to the Royal Society in that year[7]. This Kapitza–Dirac effect has been studied empirically a number of times in the last five years, a particularly complete and quantitative demonstration of its characteristics having been achieved very recently, but for a beam of sodium atoms rather than of electrons[8].

The subject of isotope separation was in everybody's mind at the Cavendish Laboratory in the early 1930s, because of the desirability of using isotopically-pure targets for the study and analysis of nuclear reaction processes. Dirac conceived the idea of separating isotopes by using the centrifugal effect, causing a jet of gas to be deflected through a large angle, in which case the atoms of the heavier species would be deflected less than those of the lighter species, leading to some degree of spatial separation between the heavy and light components. The outgoing stream could then be divided into two jets by the use of a sharp edge, the ratio heavy/light being different for the two final jets. The advantage of this scheme was that such an apparatus need have no mechanical moving parts. Kapitza encouraged Dirac to carry out the experimental work himself. R. E. Peierls was at Cambridge at that time and has written a brief account of what he saw[9]:

Kapitza allowed him the use of a compressor in the Mond Laboratory, and the device was tried initially on a mixture not of isotopes, but of air with a heavy organic compound. When I saw the experiment there had been as yet no evidence of a difference in composition between the two output tubes, but by feeling the tubes one could easily check that one was hot and the other cold, showing that something non-trivial was happening in the junction.

Dirac did not leave any detailed description of the internals of his apparatus but it appears that the centrifugal effect was obtained by means of a spiralling tube[10]. His own account of the outcome runs as follows[11]:

I got an effect which I wasn't expecting which was interesting, namely a thermal effect. I was able to produce something like a conjuring trick. I just showed three pipes, pumped in some air into one pipe; have the one pipe here branching out into two pipes and the air would come out of these two pipes at widely different temperatures ... I believe we had differences of about maybe 100° centigrade ... (It) comes entirely from the viscosity effects. The inner layers have a higher angular velocity than the outer ones and the viscosity transfers energy from the inner layers to the outer ones. The outer ones go out one side and the inner ones come out the other side and there's quite a big temperature difference.

Rutherford was highly amused by these goings-on, for he wrote as follows[12] in a letter dated 23 April 1934 thanking E. Fermi at Rome for the news he had sent of his neutron-induced nuclear reaction studies:

I congratulate you on your successful escape from the sphere of theoretical physics! You seem to have struck a good line to start with. You may be interested to hear that Professor Dirac also is doing some experiments. This seems to be a good augury for the future of theoretical physics!

However, Dirac's experiments were not completed; in particular, he did not demonstrate whether or not his apparatus could separate isotopes. Despite warnings from Rutherford and others, Kapitza made a return visit to Russia in the summer of 1934, and the Russian government did not permit him to go back to Cambridge[13]. In the absence of Kapitza, Dirac's interest in this experiment languished. In his own words[11]:

When Kapitza was detained in Russia, I rather stopped. I didn't have enough enthusiasm to carry on without him and so nothing further was done on it until the war came and people wanted to separate isotopes of uranium.

As far as we know, this was the last piece of experimental work that Dirac did.

4 Dirac's war work

The Frisch–Peierls memorandum[14] of March 1940 made it clear that the separation of uranium isotopes was going to be vital for the construction of an atomic bomb. Its authors considered at the time that this separation could be done adequately by thermal diffusion methods already known. A number of groups were set up to study possible separation methods, especially at Liverpool and Oxford. From Liverpool, M. H. L. Pryce wrote to Dirac several times in July 1940, reporting their initial discussions there and asking for some notes about Dirac's experimental work of 1934, since his method might provide an initial stage for the electromagnetic separator under consideration there[15]. We do not know Dirac's response to these letters but no work on the jet separation method was done at Liverpool, to the best of our knowledge. The Oxford group, under F. Simon[16], opted for ^{235}U separation by gaseous diffusion of uranium hexafluoride through membranes, as suggested to them by Peierls, but carried on a small research project on the jet separation method as well[10]. The Birmingham group under Peierls worked on a number of theoretical aspects of the atomic bomb project, including the ^{235}U separations methods being developed at Oxford. In the later summer of 1940, Peierls wrote an undated memorandum MS12 entitled, 'Efficiency of Isotopic Separation', discussing the general problems. This was later revised by Peierls jointly with K. Fuchs and issued[17] as memorandum MS12A, also undated, entitled 'Separation of Isotopes'. It begins with the following statement:

This Report is to replace the earlier MS12. The notation and terminology have been revised in the light of experience and several errors eliminated. Since MS12 was written we have seen a report by Dirac who comes to identical conclusions concerning the quantity we now call 'separating power' (this term is taken from Dirac) and also reports by Urey and Cohen on the same subject. These pages as well as the formulae given in a report by Simon and Kurti have been used in redrafting.

In his standard work on the theory of isotope separation[18], published in 1951, Karl Cohen gives the date of report MS12A as February 1942. We shall see below that this is consistent with other independent evidence.

The report by Dirac mentioned here is known. It bears the title 'The Theory of the Separation of Isotopes by Statistical Methods'. The only copy we know bears no date nor identifying number and lacks the Appendix referred to in the text, but it does bear a sign to indicate that it was a highly classified document. It consists of three pages of closely typed material, concerned with the general theory of those methods of separating isotopes which subject the mixture of isotopes to physical conditions, such as a centrifugal force field or a temperature gradient, which tend to cause a gradient in the concentration of the isotope. It puts an upper limit on the output of any such apparatus and shows what running conditions may allow this theoretical limit to be approached in practice. It is therefore a very fundamental contribution to this field of research. With c (assumed small) denoting the concentration of the isotope to be concentrated, α being the gradient of c when equilibrium holds for these physical conditions and β being its actual gradient, Dirac obtained the expression $\rho D\{\beta \cdot (\alpha - \beta)\}c^2$ for the diffusion flow per unit volume producing separation, where ρ is the fluid density and D its coefficient of diffusion, and introduced the name *separation-power* for this quantity, which he abbreviated as 'sep.-power'. The maximum sep.-power is reached when the apparatus is arranged such that $\beta = \frac{1}{2}\alpha$ everywhere. The 'sep.-energy' of a sample of the fluid increases with the sep.-work done on it, the latter being the time-integral of the sep.-power used. Its general form may be written as $mf(c)$ where m is the mass of the fluid and c its concentration, and Dirac obtained the following explicit expression for $f(c)$,

$$f(c) = -1 + \log(c/c_0) + (c/c_0), \tag{1}$$

where c_0 denotes the initial concentration. Expressions for the case where the concentration c is not small were apparently given in the missing Appendix. This recognition that the separative power had an upper limit, irrespective of the physical details of the procedure used, and his calculation of this theoretical maximum in an elegant and general way, constitute the most important part of Dirac's contribution to the theory of isotope separation by diffusive mechanisms.

In his book, Cohen refers[19] to this report by Dirac as 'British Report 1941'. Cohen carries the calculations out for arbitrary strength of the concentration c, 'following Dirac' as he puts it, obtaining a result which reduces to the expression (1) for small c. He uses the term 'separative power', essentially that introduced by Dirac, and this is now part of the accepted terminology, perhaps the most apparent trace today of Dirac's contribution to this field. The date of his seminal paper is not really known. It was not the response to Pryce's letters from Liverpool, for that enquiry was concerned only with centrifugal separation of a particular kind. This is a much more general paper, applicable to a wide range of diffusion separation techniques. As we have seen above, it was obviously written before the Fuchs–Peierls memorandum M 12A was finalised. Fuchs was given a qualified security clearance and signed the Official Secrets Act at the end of May 1941, and it appears reasonable to suppose that Peierls may have received this paper in the summer or autumn of 1941 and then worked through it with Fuchs. It is not yet clear how Dirac's report came to be written nor exactly when, nor is the path known by which it reached Peierls.

On 10 November 1941, Peierls wrote to Dirac as follows[20]:

I am doing some theoretical work in connection with a problem for a Government Department and we find now that there is more urgent work in connection with this problem than we can handle at Birmingham . . . There are certain aspects of the problem which I think you might be interested in, apart from the problems which we have already discussed with you some time ago.

In reply, Dirac suggested that Peierls should have supper with him on his next visit to Cambridge, on the following weekend, and it is clear that Dirac then agreed to work with the Peierls group, while remaining at home at Cambridge. The earlier problems referred to are probably to do with isotope separation by diffusion at Oxford, in view of Dirac's letter to Peierls of 29 December 1941[20]:

I should be glad to know the results of the separation experiments when they are done and would probably be able to come to Oxford to discuss them. I am afraid that I have not yet done much with the other problem,

the latter being the new problem with the Peierls group. Apparently he was given memorandum M 12A to check, for he says in a letter[20] dated 25 January 1942,

Why don't you use the abbreviation sep.-power etc.? . . . Would sep.-work be any better than sep.-energy? It is a quantity which divided by sep.-power gives a time . . .

These remarks, taken together with the foreword to MS 12A, imply that the final form of the latter was written subsequent to this letter, in accord with the date assigned to MS 12A by Cohen. It is rather likely that MS 12A was one of the reports taken to America by Peierls at the end of February 1942, a visit of about a month in which a discussion meeting with Dr. Cohen at Columbia University was an important part.

The first report written by Dirac after he began to collaborate with the Peierls group was also not concerned with the new project. On 11 May 1942, Dirac wrote to Peierls[20]:

I have written up my old work about the circulation in a self-fractionating centrifuge and have enclosed it forthwith.

I have not done any calculations on the effect of temperature variation along the tube, but I believe the effect is important and someone should work it out.

This paper[21], numbered clearly as D.1 but with 'Report Br-42' also written on it by hand, is entitled 'The Motion in a Self-Fractionating Centrifuge'. It seems probable that this study may have been carried out by Dirac some time before he took up this regular collaboration with the Peierls group, possibly in response to an enquiry by Peierls about the experimental work of Beams *et al.*[22] concerning the use of a centrifuge for isotope separation. The problem dealt with is to determine the circulation of the axial flow in a centrifuge where the initial gas is fed in at the centre of the base, the light component being extracted at the axis at the top and heavy component being removed at the periphery of the base. Dirac simplified the problem in very many ways, retaining only essential features. He assumed the rate of removal of the gas to be negligible, that the rotation of the centrifuge is so fast that essentially all of the gas is near the boundary, so that the curvature of the boundary could be neglected, and that the 'adiabatic force' term could be neglected (probably the most questionable assumption). The equations were linearized for small axial velocities and for small deviations of the azimuthal velocities from those due to the steady rotation of the centrifuge. Dirac sought eigenfunctions of the form $\phi_n(r)\exp(-k_n z)$, which fall exponentially with height z above the base of the centrifuge, and solved this problem analytically for the lowest eigenvalue k_0, which he found to be given by

$$k_0 = 1.2\mu V^5/(pa^2 R^2 T^2),$$

where V, p and a denote the velocity, pressure and radius at the periphery of the centrifuge, μ is the viscosity, R the gas constant and T the temperature of the gas. The eigenvalue k_0 specifies the eigenfunction which necessarily dominates for sufficiently large z. The characteristic length $1/k_0$ measures the rate at which the circulatory motion decreases with increasing z and so provides an estimate of the greatest axial length useful for such a centrifuge to have. In this lowest mode, the axial velocities of the gas are necessarily those of a counter-current centrifuge, the gas with the heavier fraction moving downward near its periphery and the gas with the lighter fraction moving upward further inward from the periphery, and this mode of circulation is readily set up mechanically, by a stationary scoop set up at the base of the centrifuge. In the last paragraph of his paper, Dirac notes that greater lengths could be usefully used for the centrifuge tube if the circulation were set up by a suitable temperature gradient along the tube, but he does not investigate this situation further.

This paper by Dirac was seminal in the centrifuge field. Although he did not invent the counter-current centrifuge – this was proposed by H. C. Urey in 1939 – his work provided a theoretical basis for it and drew attention to its characteristics. For isotope separation, the counter-current centrifuge has been shown to have a very high efficiency in practice, almost 100% of the maximum efficiency theoretically possible, so that it has become widely used for uranium isotope separation, for example by the British–Dutch–West German consortium URENCO. As a result, much calculational work was subsequently directed towards going beyond the simplifying assumptions made by Dirac in his early treatment. Most of this later work has been classified since about 1961 when the U.S. Government decided to investigate the counter-current centrifuge system in detail, as a possible route for the expansion of their uranium isotope separation programme, the final decision to go this way being taken in President Carter's time and now being implemented[38]. Dirac's central contribution to this field is generally recognised and frequently referred to. It is of interest to remark that Dirac's centrifuge paper is neither discussed nor referred to in Cohen's declassified text[18], although we are informed that it was discussed in full, and extended, in the much larger, classified version of this text.

The new work to be done arose out of the extensions required to the 1940 Frisch–Peierls paper[14] on energy generation in blocks of fissile material and to Peierls's 1939 paper[23] on neutron multiplication in the same. Dirac's work led to five reports on these topics. The first of these reports was MS.D.2, entitled 'Expansion of U-sphere enclosed in a container'. This was concerned with the dynamical effects of a thick container (or tamper) placed around the block of fissile material.

Without a container, a supercritical mass of uranium undergoes a rapid rise in temperature, and therefore of pressure, more or less uniformly throughout the material, because of the exponential increase of neutron multiplication and the transfer of the energy of these neutrons to the block of material as a whole. Since the external pressure on the block is only atmospheric, a rarefaction wave then moves inwards from its surface. Behind this wave, the pressure falls and the material there moves outward, its consequent fall in density leading to a cessation of neutron multiplication and energy generation there. When this wave reaches the centre of the block, the neutron multiplication stops and the available energy is concentrated at the centre of this mass. There then follows a strong outflow of energy, leading to a blast wave of high intensity.

With a container, the rapid rise of pressure in the supercritical uranium mass generates an outgoing pressure wave in the container material which can develop into a shock front. When this wave reaches the outer surface of the container, it is reflected as a rarefaction wave, which travels inward and into the uranium core. The presence of the container therefore causes delay in the expansion of the uranium mass, which means that the neutron multiplication can go on for a longer time and so lead to higher temperatures due to the increased nuclear energy release. Dirac's discussion of the dynamical aspects of this sequence of processes must have been completed early in the Summer of 1942.

The increased neutron multiplication due to the container was discussed in Dirac's next paper, numbered MS.D.3 and entitled 'Neutron Multiplication in a Sphere of Uniform Density surrounded by a Shell of Non-uniform Density'. This work was completed early in August, for Dirac wrote to Peierls as follows, in a letter dated 25 August 1942:

Thanks for the papers on shock waves . . . I was rather surprised to see how the whole theory comes out without a detailed knowledge of the processes by which the energy is degraded. I have enclosed my work on the effect of a non-uniform density in the outer layers in changing the rate of neutron multiplication.

He was much fascinated by the elegant theory G. I. Taylor had worked out[24] to describe the outgoing blast wave which was formed in the final stage of the supercritical uranium explosion, for he wrote further[20], on September 8:

Concerning the stability of Taylor's solutions, this could be investigated by considering small departures from Taylor's solutions. Do you want me to look into this?

This particular proposal was not followed up but it appears that the paper MS.D.3 had at least a substantial part in which perturbation theory methods were used for taking into account the non-uniformity of the container. As mentioned above, the container material undergoes distortion early in the sequence of processes set off by the initial neutron multiplication. Outgoing neutrons may be scattered back into the uranium core by interaction with the container nuclei and so increase the neutron multiplication further, and it is important to take this effect into account accurately in calculating the final outcome. The perturbation theory approach to estimating this effect used the rate of variation of the density with position as the expansion parameter, but it seems possible that this part of the work may have been carried out at a rather late stage in the history of this paper. There are a number of references to this aspect of the work in the later letters between Dirac and Peierls. In a letter[20] dated 1 September 1943, Dirac commented that 'there are various new terms appearing in the perturbation theory which have no analogue for the bare sphere'. On 17 October 1943, Peierls wrote[20] to Dirac to say 'I verify that your Paper on Perturbation Theory is MS.D.3', which suggests that their discussion in that period was involving a re-examination of some details related with that paper which had first been mentioned in these letters more than a year earlier. Much later still, this paper MS.D.3 is referred to again in a letter[20] of 1 May 1944 written by Peierls soon after, or during, his second visit to Los Alamos from his base in New York at that time. This letter conveys criticisms from H. A. Bethe, then at Los Alamos, who had found that the simple perturbation treatment followed by Dirac was not sufficiently accurate to describe the situation quantitatively. The assumption that the density $\rho(r)$ varied slowly with r failed because the outgoing pressure waves quickly develop into a shock wave and the derivative $\rho'(r)$ is not small in the neighbourhood of the shock front. Bethe found it was necessary to keep higher order terms giving additional contributions of the same order of magnitude as those calculated in MS.D.3.

The fourth report in this series by Dirac was MS.D.4, entitled 'Estimate of the efficiency of energy release with a non-scattering container'. On 9 October 1942, this was nearly complete, for he wrote[20]:

I have no objection to your sending any of my reports to America but it might perhaps be better to wait until the work on the efficiency is completed, as this work should not take long now and the subject will then appear in a more finished form.

At the end of September 1942, W. A. Akers, chairman of the Directorate of Tube Alloys, under whose umbrella all of these investigations were being carried out, raised the question whether Dirac's work might not be put on a more official basis, which would pave the way for a change of location for him to become more closely integrated into the Tube Alloys programme of work. Peierls did not consider this important and made only the comment that he would like to make sure that Dirac's reports were all made available to the Americans. An official appointment would have required Dirac to take leave from his professorship at Cambridge whereas he was quite able to contribute significantly to the Tube Alloys Project work while fulfilling the duties of his Chair, receiving only travelling and other minor expenses from the Government, and he certainly preferred not to leave his home at Cambridge. Nevertheless, he replied 'I do not have any preference as to whether my work is put on a more official basis or not'. Peierls found the existing situation satisfactory and Akers did not press his suggestion.

Late in 1942, before the breakdown in American–British cooperation developed, Peierls had received from J. R. Oppenheimer papers concerning the Los Alamos work on the detailed dynamics involved in the explosion of fissile material and had passed them on to Dirac for study and comment. Dirac was most concerned with the neglect of the material pressure relative to the radiation pressure and of the expansion arising from the material pressure. On 17 January 1943, he returned to Peierls a copy of Peierls's draft reply to Oppenheimer, with the remarks[20]:

I think your answer to Oppenheimer covers the whole field pretty well and gives the right amount of emphasis to each point, except that I would like to see the point about whether equilibrium is attained between matter and radiation treated more fully, as it is an important point, which they may not have considered adequately.

He also discussed a number of internal inconsistencies in the American document and also the questions how the two halves of the bomb should best be shaped and how they should be brought together for explosion, concluding that mechanical convenience would probably be the most important factor in deciding between optimal possibilities of various kinds.

The first of the two papers on the initial neutron multiplication, numbered MS.D.5 (Part I) and entitled 'Approximate Rate of Neutron Multiplication for a Solid of Arbitrary Shape and Uniform Density', was written by Dirac alone, and completed about the end of March 1943. Peierls had already laid out the essentials of the problem[23]. The time dependence of the neutron number $n(\mathbf{r}, t)$ at point $\mathbf{r}$ and time t consists of a sum of exponential terms $\exp(-\lambda_i t)$, where the λ_i are eigenvalues of a time-independent integral equation. Exponential increase of the neutron number with time will occur and lead to explosion if the lowest eigenvalue λ_0 is negative. For a small mass, the lowest eigenvalue is positive; a neutron flux in the fissile material (produced for example by incident cosmic radiation or by spontaneous fission) will then be damped out. As the fissile mass is increased, the eigenvalues change and the mass reaches its critical value when λ_0 reaches the value $\lambda_0 = 0$. In this lowest eigenstate, it is reasonable to assume that the neutron distribution is uniform over the solid. As Dirac shows, the critical condition is then determined by purely geometric considerations, since it can be obtained by quadratures from a knowledge of the distribution function $\psi(l)$ giving the relative frequency of chords connecting two points on the surface of the solid, with length between l and $(l + dl)$. This procedure is known as the method of chords. The function $\psi(l)$ satisfies a number of constraints; for example, to specify several of them, we may note that

(i) $\int \psi(l) l \, dl = V$, where V denotes the volume,

(ii) $\int \psi(l) \, dl/\pi = A$, where A denotes the surface area,

(iii) $\int \psi(l) l^4 \, dl = 12 V^2$,

the last two holding in this form only for a convex solid. The case of most interest is that of a hemi-spheroid, corresponding to one of the two pieces of fissile material to be brought together to form a spheroid of more than critical mass. The hemi-spheroid has an edge with dihedral angle $\pi/2$, and Dirac obtains the general result

(iv) $\quad \psi(0) = (4/3) \sum_i L_i \{(\pi - \gamma_i) \cot \gamma_i + 1\}$

for any solid which has edges of total length L_i with dihedral angle γ_i. For the hemi-spheroid, this expression reduces to $\psi(0) = 8\pi a/3$, where a is the end circular radius. For the oblate hemi-spheroid, where the axial length b is smaller than a, the function $\psi(l)$ has its greatest value for $l = b$, but this is a sharp maximum where the derivative $\psi'(l)$ is discontinuous, its jump at $l = b$ being calculated to be

(v) $\quad \psi'(l)|_{b+} - \psi'(l)|_{b-} = 8\pi^2 a^2/b^2.$

The interest in these and other such relationships is that a trial $\psi(l)$ satisfying them leads to a rather good approximation for the critical condition for solids which are not precisely spheroidal in shape.

The second part of this paper, numbered MS.D.5 (Part II) and entitled 'Application to the Oblate Spheroid, Hemi-sphere and Oblate Hemi-spheroid', was written jointly with K. Fuchs, R. E. Peierls and P. Preston and appears to have been completed about the end of July 1943. The chord distribution $\psi(l)$ is calculated analytically for a spheroid, and numerically for a hemisphere. The approximate methods developed in Part I are then applied to the case of an oblate hemi-spheroid; for the limiting case $b \to a$, the approximate results agree quite well with those of the numerically exact calculation for the hemisphere.

5 Dirac's jet separation method

During most of the period discussed, and later, Dirac was also in direct touch with F. E. Simon's group at Oxford. It appears that he did visit Simon at the Clarendon Laboratory early in 1942 to discuss the status of their isotope separation results, as he had offered to do in his letter of 29 December 1941 to Peierls. Although the precise date of this visit does not seem to be recorded, I am informed by N. Kurti that it was in the

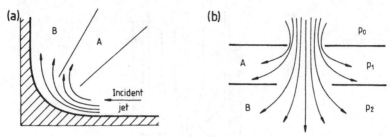

Fig. 1. Shows the two proposals which Dirac made to Simon and the Oxford group for the construction of a ^{235}U separator, using an incident jet of uranium-bearing gas. They are referred to as (*a*) the curved path slit system, and (*b*) the hole-plate system. For the latter, we use the notation *d* for the distance between the plate, and w_1 and w_2 for the diameters of the holes in the first and second plate, respectively.

course of these discussions that Dirac proposed the possibility of a simple centrifugal jet separator. His first proposal was to turn the jet by a smooth curved 90° angle, as shown on Fig. 1(*a*), dividing the final stream into two jets as shown there. Later in the discussion, he offered an even simpler system, consisting of two parallel plates with circular coaxial apertures of diameters w_1 and w_2, respectively, the set-up being as depicted on Fig. 1(*b*), the gas enriched in the light isotope being that pumped away from the region labelled A. Subsequently, J. A. V. Fairbrother undertook to supervise the project of investigating Dirac's proposals. The matter was of more than academic interest at the time since it was not certain then, nor even early in 1944, that ^{235}U separation by gaseous diffusion through membranes would necessarily be a successful project on a factory scale.

Fairbrother obtained some preliminary data with the set-up 1(*b*) for the case of an equimolar N_2–CO_2 mixture as early as October 1942, for Dirac wrote to him on 25 October to thank him for the interesting set of results he had sent. Dirac commented that, since the isotope separation increased as the initial pressure p_0 of operation was decreased, this pressure should be still further reduced, and that the pressure p_1 should always be kept much lower than p_0. On 26 November, Dirac wrote[25]:

I wonder how the separation is getting on ... There will be a transfer of energy from A to B [cf. Fig. 1(*b*)] by viscosity. I would like you to try whether this is really the case by throttling down the exit for the gas *B* and measuring exactly

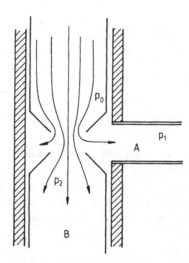

Fig. 2. Shows the most successful system used by Fairbrother and Tahourdin in their experimental work using N_2–CO_2 mixtures, the straight-through slit system developed after experience with the hole-plate system depicted in Fig. 1(*b*).

the pressure difference between p_0 and p_2 as it would be very convenient for a separation plant if the gas *B* did not have to be pumped at all.

Their correspondence continued through the winter, many sets of data being sent by Fairbrother, with replies from Dirac endeavouring to understand the effects observed – the phenomena were a function of at least five variables, the pressure p_0, the pressure ratio p_2/p_1, the gap *d*, and the slit widths w_1 and w_2, specified on Fig. 1(*b*) – and suggesting new runs which might be made to clarify the situation.

At some time before May 1943, Fairbrother was joined by an Oxford University student named P. A. I. Tahourdin, who carried out this work first as Part II (fourth year) of the B.Sc. requirements in Physical Chemistry, and later as a research student for the D.Phil. degree, for he wrote to Dirac on 17 May 1943, as follows[25]:

Please find enclosed the readings which Dr. [*sic*[26]] Tahourdin has taken with the new arrangement of slits, namely, entrance slit 0.07×3.2 mm and exit slit 0.123×3.2 mm . . . ,

including a large body of measurements. The apparatus was rebuilt and further developed to use straight-through slits (see Fig. 2), and

Fairbrother wrote on 14 October to tell Dirac that they were in business again. Dirac replied on 24 October[25]:

With the new apparatus, I would like to have the results concentrated around the optimal conditions – in particular, I would suggest that you keep to conditions for which the ratio of the flows f_2/f_1 is near unity since this ratio would have to be very close to unity in actual working conditions and the performance of the apparatus in other conditions is not of much importance or interest.

Work was continued at Oxford on the Dirac jet separation method until 1945, when Tahourdin wrote and submitted his D.Phil. dissertation[26], entitled 'The Dirac Jet Separation Method'. The straight-through slit systems proved to be the most convenient and most efficient to use, and gave increases Δc in the molar fractions of N_2 between the input N_2–CO_2 mixture and the light fraction outgoing in jet A (cf. Fig. 1) as large as 4.5% over the range of conditions investigated[10,27]. Tahourdin wrote a report on this phase of the work in August 1944, which was entitled 'The Dirac Jet Separation Method applied to Nitrogen–Carbon Dioxide Gas Mixture' and ends with the acknowledgement:

Thanks are due to Professor P. A. M. Dirac, whose ideas and suggestions have continued to form a basis of this work.

The story is taken further in Tahourdin's D.Phil. dissertation[26] in 1945. It had been expected that the curved path slit system (cf. Fig. 1(*b*)) would give a large improvement over the performance of the straight-through slit system, perhaps even by a factor of five, but the results proved to be disappointing in that the highest values measured for Δc with this system were below 2%.

Altogether, Dirac's jet separation method was not competitive with the other methods available in the 1940s and the process chosen for ^{235}U separation, gaseous diffusion through membranes, proved to be successful and reliable in the time of need. However, Dirac's work was recalled again in 1975, when reference was made in the journal *Science* to two 'new' methods for the enrichment of ^{235}U, the Becker 'separation-nozzle' in West Germany[28] and the 'helikon' technique in South Africa[29], and in the journal *New Scientist*[30] to the Becker 'nozzle enrichment process'. Following the latter, Nicholas Kurti wrote a letter of protest to the *New Scientist*[30], pointing out that the Becker nozzle process was identical with the curved-jet method invented by Dirac in 1941 and tested extensively by Tahourdin at the Clarendon Laboratory in the period 1942–5. Following the two articles in *Science*, S. A. Stern of Syracuse University similarly wrote a letter, rather longer but on the same lines, to *Science*[31] in order to draw attention to Dirac's prior invention. Apparently, technical improvements, especially in the strength of materials, have made Dirac's jet method more competitive today.

6 Conclusion

After the Quebec Agreement in late August 1943, British–American cooperation on atomic bomb development was resumed and a number of key figures in the U.K. programme joined the work in the U.S.A. Peierls went in December 1943 and Fuchs followed him in the early months of 1944. After about six months in New York, Peierls was moved on to Los Alamos since the U.S. programme on uranium isotope separation by the gaseous diffusion method appeared to be well in hand. Thus Dirac's connection with the work of the Peierls group ended late in 1943. As recounted above, his active interest in the jet method for isotope separation continued until late 1944.

One can see clearly some ambivalence in Dirac's attitudes concerning physics. On the one hand, he has repeated many, many times his conviction that beauty in the fundamental equations of physics has priority over any other considerations such as the closeness of their fit to current experiment. This may be well illustrated by a number of mathematical investigations which he made[32-35], which have not yet led to any interesting physics. Incidentally, one interesting exception to this remark is to be found in the paper[34] entitled 'Pretty Mathematics' where he records that he had found in 1927, before he discovered the Dirac equation for the electron, a relativistic two-component equation describing a neutral particle with spin $\frac{1}{2}$, the equation we associate today with the name of Majorana, that of the Majorana neutrino[36], published by Majorana in 1937. On the other hand, in the present article, we see Dirac deeply involved with approximate calculations whose only purpose was to obtain an answer which the experimenters could rely upon for the building of large and significant projects, acting effectively in the role of a consultant engineer. He also said in an interview in 1963[11]:

I owe a lot to my engineering training because it did teach me to tolerate approximations. Previously to that I thought . . . one should just concentrate on exact equations all the time. Then I got the idea that in the actual world all our equations are only approximate. We must just tend to greater and greater accuracy. In spite of the equations being approximate, they can be beautiful.

He felt that this training had given him an appreciation of how important a simple approximate formula could be. He also felt that everything might be an approximation, a feeling which he thought came very largely from his engineering training.

Dirac was clearly interested in this work he did during the War years, with an intensity going beyond what was called for by the national need. His long connection with the Oxford jet separation project suggests that he felt some considerable pleasure in the conception and investigation of a practical scheme like that.

Dirac's experimental work under Kapitza's encouragement was a more complicated matter. Dirac and Kapitza were a most unusual pair, this shy and taciturn man and that great extrovert, that Russian bear. Dirac's impulse to do that practical work in 1934 was a product of his relationship with Kapitza; his loss of interest in the project after Kapitza's confinement to Russia from 1934 on, indicates this. For the rest of his life, he felt deeply about what had been done to Kapitza. In 1934 and the years immediately following, he visited Russia whenever he could, to give encouragement to Kapitza. Even in the last days of his life, he spoke repeatedly about Kapitza in those pre-War times, both before and after Kapitza's confinement to the U.S.S.R.

Professor Paul Dirac was a deep and remarkable man, about whom we shall marvel for the rest of our days.

7 *Acknowledgements*

The account I have given in this paper of Dirac's War-time research, owes a great debt to the skill, persistence and acumen of Mr. John Clarke, Archivist with the UKAEA Archives at AERE (Harwell), in locating the relevant letter files and the declassified copies of reports, which were necessary for me to gain a full picture. I also wish to acknowledge here the willing help of Prof. Nicholas Kurti in discussing the background and origins of the isotopic separation work of the Simon group at the Clarendon Laboratory, especially concerning the work of Dr. P. Tahourdin on the Dirac separation method. Discussions with Prof. Sir Rudolf Peierls on the work of his group at Birmingham in the early years of the War and on the content of the papers prepared by Prof. Dirac (not all known to be extant, although their titles are recorded) in his War-time collaboration with the Peierls group, have been vital for this account and I acknowledge them with much gratitude.

Note added in proof

After this paper had been prepared and submitted for publication I came across a recent article by S. Whitley[39] which reviews gas centrifuge research up to 1962 and acknowledges in some detail Dirac's outstanding contributions to gas centrifuge technology.

References

1 T. Kuhn, interview with P. A. M. Dirac, 1 April 1962 – Tape 7a, Niels Bohr Library, American Institute of Physics, New York.
2 P. A. M. Dirac, *Proc. Camb. Phil. Soc.* **26**, 361 (1930).
3 See, for example, W. Heitler, *The Quantum Theory of Radiation*, 2nd edn., p. 207, Oxford Univ. Press (1944).
4 P. A. M. Dirac & J. W. Harding, *Proc. Camb. Phil. Soc.* **28**, 209 (1932).
5 For example, as in L. H. Gray's letter of 9 April 1930 to Dirac, held in the file Dirac 3/3 in the Churchill College Archives at Cambridge, concerning his measurements on photoelectric absorption coefficients with C. D. Ellis and G. T. P. Tarrant.
6 P. Kapitza & P. A. M. Dirac, *Proc. Camb. Phil. Soc.* **29**, 297 (1933).
7 D. Shoenberg, *Biographical Memoirs of Fellows of the Royal Society*, Vol. 31, pp. 347–8 (1985). The Kapitza Club ceased to exist after 1958, but was revived for this special meeting in 1966 in honour of Kapitza during his first return to Britain after he became confined to the U.S.S.R. in 1934.

8 P. L. Gould, G. A. Ruff & D. E. Pritchard, *Phys. Rev. Lett.* **56**, 827 (1986).
   The suggestion of carrying out the observations for neutral sodium atoms
   rather than electrons was made by S. Alshuler, L. M. Frantz & R.
   Braunstein, *Phys. Rev. Lett.* **17**, 231 (1966).

9 R. E. Peierls, in a letter to R. J. Eden in 1956, held at the Bodleian Library,
   Oxford, under reference No. Ms.Eng.Misc. b206.

10 P. A. I. Tahourdin, in *Final Report on the Jet Separation Method*, Oxford
   Rept. BR-694, pp. 2–3. (Clarendon Laboratory declassified version,
   29 April, 1953.) A similar helical flow device was operated in 1933 and
   described briefly by Georges Ranque in a paper entitled Expériences sur la
   détente giratoire avec production simultanées d'un échappement d'air
   chaud et d'un échappement d'air froid, *J. Phys. Rad.* **4**, 112S (1933).

11 T. Kuhn, interview with P. A. M. Dirac, 6 May 1963 – Tape 62b, Niels Bohr
   Library, American Institute of Physics, New York.

12 E. Amaldi, *Phys. Rept.* **C111**, 3 (1984).

13 L. Badash, *Kapitza, Rutherford and the Kremlin*, Yale Univ. Press, 1985.

14 O. Frisch & R. E. Peierls, Appendix 1 of ref. 37 below.

15 These two letters are held in the Dirac files in the Churchill College Archives
   at Cambridge.

16 Nancy Arms, in *A Prophet in Two Countries; the Life of F. E. Simon*, Chaps.
   10–11, Pergamon, Oxford (1966).

17 This report was declassified in 1947 and published subsequently under the
   same title as BDDA-97, HMSO, London.

18 K. Cohen, *The Theory of Isotope Separation as applied to the Large-Scale
   Production of ^{235}U*, McGraw-Hill, New York (1951).

19 *Ibid*, see pp. 20 and 125, also p. 109.

20 Dirac–Peierls correspondence (October 1941–May 1944), File LO 369(i),
   U.K.A.E.A. Archives, A.E.R.E., Harwell.

21 This report was declassified in 1946 and published subsequently under the
   same title as BDDA-7, HMSO, London.

22 J. W. Beams, *Phys. Rev.* **55**, 591(A) (1939); C. Skarstrom, H. E. Carrs &
   J. W. Beams, *ibid*.

23 R. E. Peierls, *Proc. Camb. Phil. Soc.* **35**, 610 (1939).

24 'Taylor's solution' was prompted by an argument due to G. B. Kistiakovsky
   that the energy generation by a nuclear fission chain reaction would lead
   simply to a small volume of material at a very high temperature and that this
   would be ineffective in producing a blast wave since there was no large
   volume of gas released as in a chemical explosion. Taylor disagreed with this
   last conclusion and proved his point by constructing an elegant solution for
   the outcome from a point source of energy, based on similarity arguments
   and on his knowledge of the equations governing the generation and

propagation of shock waves, which showed that about 50 % of the available energy went into the blast waves. This controversy is mentioned briefly on p. 128 of ref. 37 below. (Note added in proof: 'Taylor's solution' was published[40] soon after the War.)

25 Dirac–Fairbrother correspondence (8 December 1941–27 October 1943), File LO 571(i), U.K.A.E.A. Archives, A.E.R.E., Harwell.

26 Dr. P. A. I. Tahourdin did not supplicate for the D.Phil. degree until 30 October 1945. His D.Phil. dissertation is indexed as 'MS.D.Phil. d.480 (secret)' in the catalogue of the Radcliffe Science Library, Oxford; its title is not given there but we know it to be as given here in the text. The dissertation itself is held at the U.K.A.E.A. Archives, A.E.R.E., Harwell in Box M/805, under File LO 598. We understand that the unclassified part of this dissertation is essentially identical with ref. 10.

27 P. A. I. Tahourdin, *The Dirac Jet Separation Method applied to Nitrogen– Carbon Dioxide Gas Mixture Tube Alloys Project*, Rept. BR-418A (18 August 1944).

28 *Science* **188**, 912 (1975). The 'Nozzle' enrichment method referred to in this news item had been described much earlier by E. W. Becker, K. Bier, W. Bier, R. Schütte & D. Seidel, *Angew. Chem. Int. Ed. Engl.* **6**, 507 (1967).

29 *Science* **188**, 1090 (1975).

30 N. Kurti, *New Scientist* **65**, 533 (1975). The Monitor item which stimulated Kurti's letter was on p. 428.

31 E. C. Stern, *Science* **189**, 251 (1975).

32 P. A. M. Dirac, A positive energy relativistic wave equation, *Proc. Roy. Soc.* **A322**, 435 (1971); *ibid* **A328**, 1 (1972).

33 P. A. M. Dirac, Discrete subgroups of the Poincaré group, in *Problems of Theoretical Physics, I. E. Tamm Memorial Volume*, pp. 45–51, ed. V. I. Ritus, Nauka, Moscow (1972).

34 P. A. M. Dirac, Pretty mathematics, *Int. J. Theor. Phys.* **21**, 603 (1981).

35 P. A. M. Dirac, The future of atomic physics, *Int. J. Theor. Phys.* **23**, 677 (1984).

36 E. Majorana, *Nuovo Cim.* **14**, 171 (1937).

37 Margaret Gowing, *Britain and Atomic Energy 1939–1945*, Macmillan, London (1964).

38 D. R. Olander, *Sci. Am.* **239**, No. 2, 27 (August 1978).

39 S. Whitley, *Rev. Mod. Phys.* **56**, 41 (1984).

40 G. I. Taylor, *Proc. Roy. Soc.* **A201**, 159 (1950).

11

Playing with equations, the Dirac way[1]

A. Pais*

A great deal of my work is just playing with equations and seeing what they give.

P. A. M. Dirac

In the year 1902, the literary world witnessed the death of Zola, the birth of John Steinbeck, and the first publications of *The Hound of the Baskervilles*, *The Immoralist*, *Three Sisters*, and *The Varieties of Religious Experience*. Monet painted *Waterloo Bridge*, and Elgar composed *Pomp and Circumstance*. Caruso made his first phonograph recording and the Irish Channel was crossed for the first time by balloon. In the world of science, Heaviside postulated the Heaviside layer, Rutherford and Soddy published their transformation theory of radioactive elements, Einstein started working as a clerk in the patent office in Bern, and, on August 8, Paul Adrien Maurice Dirac was born in Bristol, one of the children of Charles Dirac, a native of Monthey in the Swiss canton of Valais, and Florence Holten. There also was an older brother, whose life ended in suicide, and a younger sister. About his father Dirac has recalled: 'My father made the rule that I should only talk to him in French. He thought it would be good for me to learn French in that way. Since I found that I couldn't express myself in French, it was better for me to stay silent than to talk in English. So I became very silent at that time – that started very early'.[3]

Dirac received his secondary education at the Merchant Venturer's School in Bristol where his father taught French. 'My father always

* Work supported in part by the Department of Energy under Contract Grant No. DE-AC02-81ER40033B.000.

encouraged me towards mathematics . . . He did not appreciate the need for social contacts. The result was that I didn't speak to anybody unless spoken to. I was very much an introvert, and I spent my time thinking about problems in nature.'[3] Throughout his life, Dirac maintained a minimal, sparse (not terse), precise, and apoetically elegant style of speech and writing. Sample: his comment on the novel *Crime and Punishment*: 'It is nice, but in one of the chapters the author made a mistake. He describes the sun as rising twice on the same day.'[4]

Dirac completed high school at age 16, then went to the University of Bristol where he graduated in 1921 with a degree in engineering. He stayed on for further study in pure and applied mathematics until, in the autumn of 1923, he enrolled at Cambridge where, nine years later, he would succeed Larmor to the Lucasian Chair of Mathematics once held by Newton.[5] It was Fowler who, in Cambridge, introduced Dirac to the old quantum theory, and it was from him that he first learned of the atom of Rutherford, Bohr, and Sommerfeld.

Dirac first met Bohr in May 1925 when the latter gave a talk in Cambridge on the fundamental problems and difficulties of the quantum theory. Of that occasion Dirac said later: 'People were pretty well spellbound by what Bohr said . . . While I was very much impressed by (him), his arguments were mainly of a qualitative nature, and I was not able to really pinpoint the facts behind them. What I wanted was statements which could be expressed in terms of equations, and Bohr's work very seldom provided such statements. I am really not sure how much my later work was influenced by these lectures of Bohr's . . . He certainly did not have a direct influence, because he did not stimulate one to think of new equations'.[6]

A few months later, Heisenberg's first paper on quantum mechanics came out. 'I learned about this theory of Heisenberg in September, and it was very difficult for me to appreciate it at first. It took two weeks; then I suddenly realized that the noncommutation was actually the most important idea that was introduced by Heisenberg'.[7] The result was Dirac's first paper on quantum mechanics[8] containing the relation $pq - qp = h/2\pi i$, independently derived shortly before by Born and Jordan. The respective authors were unaware of one another's results. Born has described his reaction upon receiving Dirac's paper: 'This was – I remember well – one of the greatest surprises of my scientific life. For the name Dirac was completely unknown to me, the author appeared to be a youngster, yet everything was perfect in its way and admirable'.[9]

In those days, Dirac invented several notations which are now part of our language: q-numbers, where 'q stands for quantum or maybe queer'; c-numbers, where 'c stands for classical or maybe commuting'.[10] He has described his work habits in those years: 'Intense thinking about those problems during the week and relaxing on Sunday, going for a walk in the country alone'.[10] (Dirac also liked mountains. Later he climbed the Elbruz in the Caucasus, together with Igor Tamm.)

In May 1926, Dirac received his Ph.D. on a thesis entitled 'Quantum Mechanics'.[11] Meanwhile, Schroedinger's papers on wave mechanics had appeared, to which Dirac reacted with initial hostility, then with enthusiasm. He quickly applied the theory to systems of identical particles.[12] At almost the same time, that problem also attracted Heisenberg,[13] whose main focus, on few particle systems, resulted in his theory of the helium atom.[14] Dirac's paper[12] (August 1926), on the other hand, will be remembered as the first in which quantum mechanics is brought to bear on statistical mechanics. Recall that the earliest work on quantum statistics, by Bose and by Einstein, predates quantum mechanics. Also, Fermi's introduction of the exclusion principle in statistical problems, though published[15] after the arrival of quantum mechanics, is still executed in the context of the 'old' quantum theory.[16] All these contributions were given their quantum mechanical underpinnings by Dirac who was, in fact, the first to give the correct justification of Planck's law, which started it all: 'Symmetrical eigenfunctions ... give just the Einstein–Bose statistical mechanics ... (which) leads to Planck's law of black-body radiation.'[12]

It is edifying to remember that it took some time before it was sorted out when Bose–Einstein and Fermi–Dirac statistics respectively apply. Dirac in August 1926: 'The solution with anti-symmetric eigenfunctions (F.D. statistics) ... is probably the correct one for gas molecules, since it is known to be the correct one for electrons in an atom, and one would expect molecules to resemble electrons more closely than light-quanta.'[12] Other great men were not at once clear either about this issue, Einstein, Fermi, Heisenberg, and Pauli among them.

In September, Dirac went to Copenhagen. 'I admired Bohr very much. We had long talks together, long talks in which Bohr did practically all the talking.'[6] It was there that he worked out the theory of canonical transformations in quantum mechanics since known as the transformation theory.[17] 'I think that is the piece of work which has most pleased me of all the works that I've done in my life ... The transformation theory (became) my darling.'[7] In this paper, Dirac introduced an important tool of modern physics, the δ-function, about which he remarked right away: 'Strictly, of course, $\delta(x)$ is not a proper function of x, but can be regarded as the limit of a certain sequence of functions. All the same, one can use $\delta(x)$ as though it were a proper function for practically all the purposes of quantum mechanics without getting incorrect results'.[18]

Dirac's stay in Copenhagen – lasting till February 1927 – is also highly memorable, because it was there that he completed the first[19] of two papers in which he laid the foundations of quantum electrodynamics. The sequel[20] was written in Goettingen, the next important stop on his journey.

Preceding these two papers, Dirac had already given[12] a theory of induced radiative transitions by treating atoms quantum mechanically but still considering the Maxwell field as a classical system.[22] However, 'one cannot take spontaneous emission into account without a more elaborate theory'.[12] Here, Dirac echoed Einstein who, already in 1917, still the days of the old quantum theory, had stressed that spontaneous emission 'make(s) it almost inevitable to formulate a truly quantized theory of radiation.'[22] In his Copenhagen paper,[19] Dirac did just that. He proceeded to quantize the electromagnetic field, thereby giving the first rational description of light quanta, and then derived from first principles Einstein's phenomenological coefficient of spontaneous emission.[23]

The theory was not yet complete, however: 'Radiative processes ... in which more than one light quantum take(s) part simultaneously are not allowed on the present theory.'[19] How young quantum mechanics still was. Early in 1927, Dirac did not yet know that these processes are perfectly well included in his theory. All one had to do was extend perturbation theory from first order (used by him in the treatment of spontaneous emission) to second order. So, in his Goettingen paper,[20] he developed[24] second order perturbation theory, which enabled him to give the quantum theory of dispersion.[25] He further noted[26] that the theory could now also be applied to the Compton effect, a subject that had interested him earlier.[27]

From Goettingen, Dirac went to Leiden and concluded his travels by attending the Solvay conference in Brussels (in October), where he met Einstein for the first time. From discussions with Dirac, I know that he admired Einstein. The respect was mutual. ('. . . Dirac to whom, in my opinion, we owe the most logically perfect presentation of (quantum mechanics)'.[28]) Yet, the contact between the two men remained minimal. I do not believe that this is due to Einstein's critical attitude to quantum mechanics, expressed first at that 1927 Solvay conference. Indeed, as time went by, Dirac himself developed reservations not only regarding quantum field theory but also, though less strongly, in relation to ordinary quantum mechanics.[29,30] I would rather think that it was not in Dirac's personality to seek for father figures.

Dirac has recalled a conversation with Bohr during the 1927 Solvay conference. Bohr: 'What are you working on?' Dirac: 'I'm trying to get a relativistic theory of the electron.' Bohr: 'But Klein has already solved that problem'.[6]

Dirac disagreed.

By the time of the 1927 Solvay conference, a relativistic wave equation was already known: the scalar wave equation, stated independently by at least six authors,[31] Klein and Schroedinger among them. One could not, it seemed, associate a positive definite probability with that equation, however. That Dirac did not like at all, since the existence of such a density was (and is) central to his transformation theory. 'The transformation theory had become my darling. I was not interested in considering any theory which would not fit in with my darling . . . I just couldn't face giving up the transformation theory.[7] That is why, as said, Dirac disagreed with Bohr. Accordingly, he began his own search for a relativistic wave equation that does have an associated positive probability density. Not only did he find it but, in the course of doing so, he also discovered the relativistic quantum mechanical treatment of spin.

That was a major novelty. In May 1927, Pauli had proposed[32] that the electron satisfy a two-component wave equation which does contain the electron spin, explicitly coupled to the electron's orbital angular momentum. Nothing determined the strength of that coupling, the 'Thomas factor', which had to be inserted by hand 'without further justification'. This flaw, Pauli noted, was due to the fact that his equation did not satisfy the requirements of relativity. The theory was, in his words, provisional and approximate.

In his equation, Pauli described the spin by 2×2 matrices, since known as the Pauli matrices. It appears that Dirac had discovered these independently: 'I believe I got these (matrices) independently of Pauli, and possibly Pauli also got them independently from me.'[6] Always in quest of a relativistic wave equation with positive probability density, Dirac continued playing[33] with the spin matrices. 'It took me quite a while ... before I suddenly realized that there was no need to stick to quantities ... with just two rows and columns. Why not go to four rows and columns'.[6] Quite a while, actually, was only a few weeks. Toward the end of his life, Dirac reminisced: 'In retrospect, it seems strange that one can be so much held up over such an elementary point (!)'.[34]

Thus, early in 1928, was born the Dirac equation[35,36] with the positive density its author had so fervently desired. To his great surprise, he had stumbled on much more, however.

It was found that this equation gave the particle a spin of half a quantum. And also gave it a magnetic moment. It gave just the properties that one needed for an electron. That was really an unexpected bonus for me, completely unexpected.[7]

Spin was a necessary consequence, the magnetic moment and the Sommerfeld fine structure formula came out right, the Thomas factor appeared automatically; and for energies small compared to mc^2 (m = electron mass) all the results of the nonrelativistic Schroedinger theory were recovered. Dirac had played hard and played well. His discovery ('once you got the right road it jumps at you without any effort'[37]), ranking as it does among the highest achievements of twentieth century science, is all the more remarkable since it was made in pursuit of what eventually turned out to be a side issue, positive probabilities.[38]

Along with its spectacular successes, the Dirac equation was, for a few years, also a source of great trouble, however.

Pauli's wave functions have two components, corresponding to the options spin up and spin down. But Dirac's wave functions had four. The question: why four? led to monumental confusion about which, in the 1960s, Heisenberg recalled: 'Up till that time (1928), I had the impression that, in quantum theory, we had come back into the harbor, into the port. Dirac's paper threw us out into the sea again.'[39]

From the outset,[35] Dirac had correctly diagnosed the cause for this doubling of the number of components. There are two with positive, two with negative energies, each pair with spin up/down. What to do with the negative energy solutions? 'One gets over the difficulty on the classical theory by arbitrarily excluding those solutions that have a negative energy. One cannot do this in the quantum theory, since, in general, a perturbation will cause transitions from states with E positive to states with E negative.'[35] He went on to speculate that negative energy solutions may be associated with particles whose charge is opposite to that of the electron. In that regard, Dirac did not yet know as clearly what he was talking about as he would one and a half years later. This undeveloped idea led him to take the problem lightly, initially: 'Half of the solutions must be rejected as referring to the charge $+e$ of the electron'.[35] In a talk given in Leipzig, in June 1928, he no longer spoke of rejection, however. Transitions to negative energy states simply could not be ignored. 'Consequently, the present theory is an approximation.'[40]

While in Leipzig, Dirac, of course, visited Heisenberg (recently appointed there), who must have been well aware of these difficulties. In May, he had written to Pauli: 'In order not to be forever irritated with Dirac, I have done something else for a change',[41] the something else being his quantum theory of ferromagnetism. Dirac and Heisenberg discussed several aspects of the new theory.[42] Shortly thereafter, Heisenberg wrote again to Pauli: 'The saddest chapter of modern physics is and remains the Dirac theory',[43] mentioned some of his own work, which demonstrated the difficulties, and added that the magnetic electron had made Jordan *trübsinnig* (melancholic). At about the same time, Dirac, not feeling so good either, wrote to Klein: 'I have not met with any success in my attempts to solve the $\pm e$ difficulty. Heisenberg (whom I met in Leipzig) thinks the problem will not be solved until one has a theory of the proton and electron together'.[44]

Early in 1929, both Dirac and Heisenberg made their first trip to the United States, Dirac lecturing at the University of Wisconsin, Heisenberg at the University of Chicago. In August of that year, the two men boarded ship together in San Francisco, stopped over in Hawaii,[45] then went on to Japan where they both lectured in Tokyo and Kyoto. I was curious whether they had discussed the problematics of the Dirac equation during their trip, so I asked Dirac. He replied: 'In 1929, Heisenberg and I crossed the Pacific and spent some time in Japan together. But we did not have any technical discussions together. We both just wanted a holiday and to get away from physics. We had no discussions of physics, except when we gave lectures in Japan and each of us attended the lectures of the other. I do not remember what was said on these occasions, but I believe there was essential agreement between us.'[46]

In the meantime, Weyl had made[47] a new suggestion regarding the extra two components: 'It is plausible to anticipate that, of the two pairs of components of the Dirac quantity, one belongs to the electron, one to the proton.' In December 1929, Dirac (back in Cambridge) dissented[48]: 'One cannot simply assert that a negative energy electron *is* a proton', since this would violate charge conservation if an electron jumps from a positive to a negative energy state.[49] Rather, 'Let us assume . . . that all the states of negative energy are occupied, except, perhaps, for a few of very small velocity', this occupation being one electron per state, as the exclusion principle demands. Imagine that one such negative energy electron is removed, leaving a hole in the initial distribution. The result is a rise in energy and in charge by one unit. This hole, Dirac noted, acts like a particle with positive energy and positive charge: 'We are . . . led to the assumption that the holes in the distribution of negative energy electrons are the protons'.[49]

The identification of holes with particles is fine, but why protons? Dirac later remarked: 'At that time . . . everyone felt pretty sure that the electrons and the protons were the only elementary particles in Nature'.[50] (Recall that, in 1929, the atomic nucleus was still believed to be built up of protons and electrons![51])

Just prior to submitting his paper, Dirac wrote a letter[52] to Bohr which shows that he knew quite well that, at least in the absence of interactions, his holes should have the same mass as the electrons themselves. It was his hope (an idle one) that this equality would be violated by electromagnetic interactions: 'So long as one neglects interaction, one has complete symmetry between electrons and protons; one could regard the protons as the real particles and the electrons as the holes in the distribution of protons of negative energy. However, when the interaction between the electrons is taken into account, this symmetry is spoilt. I have not yet worked out mathematically the consequences of the interaction . . . One can hope, however, that a proper theory of this will enable one to calculate the ratio of the masses of proton and electron'. Actually the 'complete symmetry' of which Dirac wrote, charge conjugation invariance, extends to the electromagnetic interactions as well. For want of a better procedure, Dirac briefly considered the mass m in his equation to be the average of the proton and the electron mass.[53]

The confusion lasted all through 1930 when first Oppenheimer[54], then, independently, Tamm[55] noted that the proton proposal would make all atoms unstable because of the process: proton + electron → photons. In November 1930, Weyl took a new stand[56] in regard to the protons: 'However attractive this idea may seem at first, it is certainly impossible to hold without introducing other profound modifications ... indeed, according to (the hole theory), the mass of the proton should be the same as the mass of the electron; furthermore ... this hypothesis leads to the essential equivalence of positive and negative electricity under all circumstances ... the dissimilarity of the two kinds of electricity thus seems to hide a secret of Nature which lies yet deeper than the dissimilarity between past and future ... I fear that the clouds hanging over this part of the subject will roll together to form a new crisis in quantum physics'.

Then, in May 1931, Dirac bit the bullet[57] (or, in his words, he made 'a small step forward'[37]): 'A hole, if there were one, would be a new kind of particle, unknown to experimental physics, having the same mass and opposite charge of the electron.' Dirac initially called the new particle anti-electron. Just before the year's end, Carl Anderson made the first announcement[58] of experimental evidence for the anti-electron. The name positron first appeared in print in one of his later papers.[59] The prediction and subsequent discovery of the positron rank among the great triumphs of modern physics.

That, however, was not at once obvious.

The detection of the positron was considered by nearly everyone as a vindication of Dirac's theory. Yet its basic idea, a positron is a hole in an infinite sea of negative electrons, remained unpalatable to some, and not without reason. Even the simplest state, the vacuum, was a complex consisting of infinitely many particles, the total filled sea. Interactions between these particles left aside, the vacuum had a negative infinite 'zero point energy' and an infinite 'zero point charge'. Pauli did not like that. Even after the positron had been discovered, he wrote to Dirac: 'I do not believe in your perception of "holes" even if the "anti-electron" is proved'.[60] That was not all, however. Pauli to Heisenberg one month later: 'I do not believe in the hole theory, since I would like to have asymmetries between positive and negative electricity in the laws of nature (it does not satisfy me to shift the empirically established asymmetry to one of the initial state)'.[61]

The zero point energy and charge are actually innocuous and can be eliminated by a simple reformulation of the theory.[62] Even thereafter, the theory is still riddled with infinities caused by interactions, however. To this day, the influence of interactions cannot be treated rigorously. Rather, one uses the fact that the fundamental charge e is small, more precisely that the dimensionless number $\alpha = e^2 hc \simeq 1/137$ is small, and expands in α. To leading power in α, theoretical predictions were excellent for processes like photo-electron scattering, the creation and annihilation of electron–positron pairs, and many others. Contributions to these same processes stemming from higher powers in α are invariably infinitely large, however. One was faced with a crisis: how to cope with a theory which works very well approximately but which makes no sense rigorously. As Pauli put it in 1936 during a seminar given in Princeton: 'Success seems to have been on the side of Dirac rather than of logic'.[63] Or, as Heisenberg put it[64] in a letter to Pauli (1935): 'In regard to quantum electrodynamics, we are still at the stage in which we were in 1922 with regard to quantum mechanics. We know that everything is wrong. But, in order to find the direction in which we should depart from what prevails, we must know the consequences of the prevailing formalism much better than we do.' Heisenberg was, in fact, one of that quite small band of theoretical physicists who had the courage to explore those aspects of quantum electrodynamics which remained in an uncertain state until the late 1940s, when renormalization would provide more systematic and more successful ways of handling the problems.

The first steps toward renormalization go back once again to Dirac. In August 1933, he had written[65] to Bohr: 'Peierls and I have been looking into the question of the change in the distribution of negative energy electrons produced by a static electric field. We find that this changed distribution causes a partial neutralization of the charge producing the field ... If we neglect the disturbance that the field produces in negative energy electrons with energies less than $-137mc^2$, then the neutralization of charge produced by the other negative energy electrons is small and of the order 136/137 ... The effective charges are what one measures in all low-energy experiments, and the experimentally determined value of *e* must be effective charge on an electron, the real value being slightly bigger ... One would expect some small alterations in the Rutherford scattering formula, the Klein–Nishina formula, the Sommerfeld fine structure formula, etc., when energies of the order mc^2 come into play.'

Transcribed in the modern vernacular, Dirac's effective charge is our physical charge; his real charge our bare charge; his neutralization of charge our charge renormalization; and his disturbance that the field produces in negative energy electrons our vacuum polarization.

In quantitative form, the results Dirac had mentioned to Bohr are first found in his report[67] to the seventh Solvay conference (October 1933), the paper that marks the beginning of positron theory as a serious discipline. There, Dirac also gives the finite contribution to the vacuum polarization[68] which, in 1935, was to be evaluated by Uehling[69] for an electron moving in a hydrogen-like atom – a result which, in turn, was to provide the direct stimulus for the celebrated Lamb shift experiments of 1946.

With Dirac's Solvay report his exquisite burst of creativity at the outer frontiers of physics, spanning eight years, comes to an end.

The years 1925–33 are the heroic period in Dirac's life, during which he emerged as one of the principal figures in twentieth century science and changed the face of physics. He himself has called those years in his scientific career 'the exciting era'.[70] My foregoing sketch of that period is not, by any means, complete. For example, in 1931, Dirac produced[57] the first application of global topology to physics, his proof that the existence of magnetic monopoles implies, quantum mechanically, that electric charge is quantized. He returned to this subject some twenty years later[71] (he lectured upon it[72] at the Pocono conference, March 31–April 1, 1948) and, once again, nearly thirty years thereafter.[73] As these intervals illustrate, Dirac remained scientifically active for the fifty years following the developments that came to a close in 1933.

Rather than discussing in detail Dirac's work during that half century, I shall confine myself to brief indications of his main pursuits, in which he continued to show his high mathematical inventivity and craftsmanship but no longer that almost startling combination of novelty and simplicity that mark his heroic period. I shall further add some comments on his style; and some personal reminiscences.

Without pretence to completeness, and in fairly random order, here are some main themes which, as I see it, convey the flavor of his thinking in his later years.

Elaborations of Hamiltonian dynamics. These include studies of the special relativistic dynamical evolution of systems on various types of hypersurfaces, in classical theory[74] and in quantum mechanics.[75] Also, investigations of constrained Hamiltonian systems,[76,77] leading to his Hamiltonian formulation of general relativity.[78] That work, in turn, aroused his interest in gravitational waves.[79] Did Dirac coin the name graviton? According to the *New York Times* of January 31, 1959, 'Professor Dirac proposed that the gravitational wave units be called gravitons.'

Related to Dirac's lifelong interest in general relativity are his papers on wave equations in conformal,[80] de Sitter,[81] and Riemannian spaces.[82] He lectured on general relativity until in his seventies.[83]

Cosmological issues. In these he had become interested already in his Goettingen days.[84] He did not publish on this subject until 1937.[85] From then on, until the end of his life, he was much intrigued by the possibility that the fundamental constants in nature actually are not constant but depend on time in a scale set by the cosmological epoch, the time interval between the big bang and the present.[86] It was his hope that simple relations should emerge between such extremely large but roughly comparable numbers as the ratio of epoch to atomic time intervals and the ratio of electric to gravitational forces between an electron and a proton.[87] No definitive advance was ever achieved. Others followed these exploits with more interest than enthusiasm.

The aether. A brief period (1951–3) of speculations to the effect that quantum mechanics allows for the existence of an aether.[88]

Quantum electrodynamics. One further contribution still belongs to the heroic period. In March 1932, Dirac proposed a 'many time formalism' in which an individual time is assigned to each electron.[89] This new version of the theory, equivalent to earlier formulations,[90] marks an important first step toward the manifestly covariant procedures that were to play such a key role from the late 1940s on.

A few years later, Dirac turned highly critical of quantum electrodynamics. On the one hand, the work he produced as a result of this negative attitude has not in any way enhanced our understanding of fundamental issues. On the other hand, these later struggles are of prime importance for an understanding of Dirac himself. His radically modified position resulted from his work[67] on vacuum polarization in which he had encountered the infinities that, as said, constituted a crisis in the quantum field theory of the 1930s.

Dirac's drastic change in attitude is starkly expressed in a brief paper he wrote in 1936, his first publication following his involvement[67] with the implications of positron theory. I regard it as significant that this article followed a period during which he had not published at all for more than a year. The apropos was a fleeting experimental doubt about the validity of the theory of photon–electron scattering. Dirac reacted[91] as follows: 'The only important part (of theoretical physics) that we have to give up is quantum electrodynamics ... we may give it up without regrets ... in fact, on account of its extreme complexity, most physicists will be very glad to see the end of it.'

At this point, it should be recalled that the germs of the difficulties with the infinities date back to the classical era. A classical electron considered as a point particle has an infinite energy due to the coupling to its own electrostatic field. With this in mind, Dirac adopted the strategy of attempting to modify the classical theory first, so as to rid it of *its* infinities, and thereupon to revisit the quantum theory in the hope that also there all would now be well. At that time, that approach was followed also by others, Born, Kramers, and Wentzel among them. Even today, there remains a much needed understanding of what lies beyond the infinities. There are overwhelming reasons, however, why a return to the classical theory is the wrong way to go.[92,93]

Be that as it may, Dirac tried several times to reformulate the classical theory of the electron. His first attempt[94] dates from 1938. 'A new physical theory is needed which should be intelligible both in the classical and in the quantum theory and our easiest path of approach is to keep within the confines of the classical theory.' He started from the observations that Lorentz' classical theory of the electron's motion is not rigorously valid for high accelerations, since Lorentz' electron has a finite radius. Dirac, instead, started from a zero radius electron and was able to find a rigorous classical equation of motion for it which is free of the classical infinities but which exhibits new pathologies: it has solutions corresponding to accelerations even in the absence of external fields. He did find a not very palatable constraint that eliminates these unwanted solutions – but there was more trouble. New infinities arose upon quantizing the theory.[95] In order to eliminate these, Dirac introduced[96] what amounts to photons of negative energy. He attempted to eliminate the physical paradoxes resulting from this new postulate by introducing an indefinite metric in Hilbert space.[97] That, however, leads to still further difficulties, critically analyzed by Pauli.[98] These new postulates were never discussed in the context of positron theory.

Unable to find a satisfactory quantum version of his point electron, Dirac never mentioned this theory again in later years. By 1946, he tended to the view that the infinities are a mathematical artifact resulting from expansions in α that are actually invalid.[99]

Shortly thereafter, in the years 1947–8, quantum electrodynamics took a new turn when the renormalization program was systematically developed. That technique does not fully resolve the problem of the infinities. The electron's mass and charge unalterably remain infinite. To a very large extent, these two infinities can be rendered harmless, however, in the sense that predictions to arbitrarily high orders in α can now be made for the scattering, creation, and annihilation processes mentioned earlier, where, before, the leading order in α had worked so well, but the higher orders had been intractable. As a result, quantum electrodynamics could now be confronted with experiment to vastly improved orders of magnitude. The results were spectacular. With good reason, Feynman has called[100] the new version of quantum electrodynamics 'the jewel of physics – our proudest possession.'

Dirac would have none of it.

In 1951, he wrote: 'Recent work by Lamb, Schwinger and Feynman and others has been very successful ... but the resulting theory is an ugly and incomplete one.'[101] He had a deep aversion to the way infinite masses and charges are manipulated in the renormalization program. In that year, he started all over again for a second time in his search for a new, classical, point of departure. 'The troubles ... should be ascribed ... to our working from the wrong classical theory.'[101] His new suggestion may be considered as the extreme opposite of what he had proposed in 1938. This time, he began with a classical theory that does not contain discrete particles at all. 'The theory of electrons should be built up from a classical theory of the motion of a continuous stream of electricity[102] rather than the motion of point charges. One then looks upon the discrete electrons as a quantum phenomenon.'[103] After 1954, this model, too, vanished from his writings without leaving a trace.

Thus, from the early 1950s on, Dirac went his own lonely way. He accepted the successes of the renormalization method. In fact, in the mid-'60s, he lectured on the anomalous magnetic moment and Lamb shift calculations.[104] He never wavered in his belief, however, that quantum electrodynamics needed a new starting point. In later years, he would occasionally seek new remedies in a reformulation not so much of classical as of quantum theory.[105] In 1970, he invented the last of the Dirac equations, a relativistic wave equation with positive energies only.[106]

In September 1971, at age 69, Dirac started a new phase of his career: Professor of Physics at the Florida State University in Tallahassee. It was to become a happy period in his life during which he wrote another forty papers. The last of these (1984), entitled 'The inadequacies of quantum field theory,'[107] contains his last judgement on quantum electrodynamics: 'These rules of renormalization give surprisingly, excessively good agreement with experiments. Most physicists say that these working rules are, therefore, correct. I feel that is not an adequate reason. Just because the results happen to be in agreement with observation does not prove that one's theory is correct.' The paper concludes with Dirac's final published scientific words:

I have spent many years searching for a Hamiltonian to bring into the theory and have not yet found it. I shall continue to work on it as long as I can, and other people, I hope, will follow along such lines.

Dirac died on October 20, 1984, aged 82. He was buried in the Roselawn cemetery in Tallahassee rather than at Westminster Abbey, as members of the Order of Merit traditionally are. It was his family's wish that he should rest where he left the world.

The first edition of Dirac's book, *Quantum Mechanics*, has been on my shelves since my graduate student days in Holland. Learning from it the beauty and power of that compact little Dirac equation was a thrill I shall never forget. Years later, in January 1946, I first met Dirac and his wife on a brief visit to their home in Cambridge. I saw much more of him in the autumn of that year when we met at the Institute for Advanced Study in Princeton.[108] We would often have lunch together. It was on one of those occasions that I had my first exposure to the Dirac style of exhaustive inquiry. Because of a large appetite and a Dutch background, I would regularly eat three sandwiches at that time. One day, Dirac queried me. (Between each answer and the next question there was a half minute's pause.) D. Do you always eat three sandwiches for lunch? P. Yes. D. Do you always eat the same three sandwiches for lunch? P. No, it depends on my taste of the day. D. Do you eat your sandwiches in some fixed order? P. No. Some months later, when a young man named Salam visited me at the Institute, he said: I have regards for you from Professor Dirac in Cambridge. He wants to know if you still eat three sandwiches for lunch. Dirac and I again often lunched together when he came back to the Institute for academic 1947-8. On an early occasion, Dirac looked at my plate and noted, triumphantly: 'Now you only eat two sandwiches for lunch.' Another recollection: A corridor conversation at the Institute. D. My wife wants to know if you can come for dinner tonight. P. I regret. I have another engagement. D. Goodbye. Nothing unfriendly implied. Nothing else said like 'Some other time perhaps.' The question had been posed and answered, the conversation was finished.

Everything had been arranged at the Institute for Dirac's next visit in academic 1954-5. It was not to be. The events of the troubled spring of 1954 were summarized in the News and Views column of *Physics Today*, July 1954, under two headings: The Oppenheimer Case; Dirac denied Visa. Dirac had been informed by the American Consulate in London that he was ineligible for a visa under Section 212A of the Immigration and Naturalization Act, the infamous McCarran Act which (to quote *Physics Today*) 'Covers categories of undesirables ranging from vagrants to stowaways.' The reasons for this decision have never become quite clear, but it was believed that Dirac's visits to Russia (for scientific purposes) had much to do with it.[109] The event, widely reported in the press,[110] caused some American physicists to write to the *New York Times*: 'If this is what the McCarran Act means in practice, it seems to us a form of cultural suicide.'[111] It was a quite bad, yet by no means the worst, case of harm done during that period. It passed; Dirac was to spend two more academic years in Princeton.[108]

During all these visits to the Institute and also in the course of encounters elsewhere, I came to know Dirac quite well. A friendship developed. In the course of joint talks and walks and wood chopping expeditions, I developed a good grasp of his views on physics. Time and again, I would draw him out about his discontent with quantum electrodynamics. He would concede the successes of renormalization but forever was of the opinion that the remaining mass and charge infinities 'ought not to be there. They remove them artificially.'[107] This diagnosis may well be much better than the cures he proposed.

Other recollections: his evident pride at having invented the bra and ket notations, announced in a paper[112] specially written for that purpose; his reply to my question, posed in the early 1960s, why space reflexion and time reversal invariance do not appear in his book on quantum mechanics: 'Because I did not believe in them.' Indeed, in 1949, he had written: 'I do not believe there is any need for physical laws to be invariant under these reflections, although all the exact laws of nature so far known do have this invariance.'[113]

By far, the most revealing insight I gained from those discussions concerned the Dirac way of playing with equations, which can be summed up like this: First play with pretty mathematics for its own sake, then see whether this leads to new physics.

Throughout most of his life, that attitude is manifest in his writings. At age 28: 'There are, at present, fundamental problems in theoretical physics ... the solution of which ... will presumably require a more drastic revision of our fundamental concepts than any that have gone before. Quite likely, these changes will be so great that it will be beyond the power of human intelligence to get the necessary new ideas by direct attempts to formulate the experimental data in mathematical terms. The theoretical worker in the future will, therefore, have to proceed in a more direct way. The most powerful method of advance that can be suggested at present is to employ all the resources of pure mathematics in attempts to perfect and generalize the mathematical formalism that forms the existing basis of theoretical physics, and after each success in this direction, to try to interpret the new mathematical features in terms of physical entities,'[57] just what is happening these days, in 1986. At age

36: 'As time goes on, it becomes increasingly evident that the rules which the mathematician finds interesting are the same as those which Nature has chosen.'[29] At age 60: 'I think it's a peculiarity of myself that I like to play about with equations, just looking for beautiful mathematical relations which maybe don't have any physical meaning at all. Sometimes they do.'[2] At age 78: 'A good deal of my research work in physics has consisted in not setting out to solve some particular problem, but simply examining mathematical quantities of a kind that physicists use and trying to fit them together in an interesting way, regardless of any application that the work may have. It is simply a search for pretty mathematics. It may turn out later that the work does have an application. Then one has good luck.'[34] In that last paper, he gave three examples of the way he played: the Dirac equation, monopoles, and the last Dirac equation.[106] His own judgement, at age 69: 'My own contributions since (the) early days have been of minor importance.'[114]

What kinds of mathematics did Dirac consider pretty? 'The research worker, in his efforts to express the fundamental laws of Nature in mathematical form, should strive mainly for mathematical beauty. He should take simplicity into consideration in a subordinate way to beauty ... It often happens that the requirements of simplicity and beauty are the same, but where they clash the latter must take precedence.'[29] It is, of course, idle to argue about such subjective issues as the distinction between beauty and simplicity.

Dirac was a very private man, not much given to reminiscing about other personalities or past events. He would only rarely talk about himself. I, therefore, never felt close to having a rounded picture of his personality. In particular, much about the years of his youth has remained obscure to me. On a few occasions, he would reveal some of his emotions in his writings, however. I find it striking that, as mentioned, he would refer to the transformation theory as 'my darling'.[7] Equally notable are his rare utterances about anxiety. When, at age 60, he was asked about his feelings on discovering the Dirac equation he replied: 'Well, in the first place, it leads to great anxiety as to whether it's going to be correct or not ... I expect that's the dominating feeling. It gets to be rather a fever ...'[2] At age 67: 'Hopes are always accompanied by fears, and, in scientific research, the fears are liable to become dominant.'[115]

At age 69: 'I think it is a general rule that the originator of a new idea is not the most suitable person to develop it, because his fears of something going wrong are really too strong...'[114]

Finally, two Dirac stories.

Once Dirac and E. M. Forster met at a dinner in Cambridge. The tale of the exchange between them, in its entirety, has often been told. D. What happened in the cave? F. I don't know. That version is apocryphal, however. Peierls has told me that he asked Dirac what really was said, and got this answer. D. Was there a third person in that cave? F. No.

The other story is not about Dirac, but one that I have heard Dirac tell more than once, with relish. In a small village, a newly appointed priest went to call on his parishioners. On a visit to a quite modest home, he was received by the lady of the house. He could not fail to notice that the place was teeming with children and asked her how many the couple had. Ten, she replied, five pairs of twins. Astonished, the priest asked: You mean you always had twins? To which the woman replied: No Father, sometimes we had nothing. Precision at that level had an immense appeal to Dirac.

As I look back on the almost forty years I knew Dirac, all memories are fond ones. In some, but only some, ways he reminds me of Einstein: one of the century's great contributors, always going his own way, not making a school, compelled by the need for beauty and simplicity in physical theory, in his later years, more addicted to mathematics than was good for his physics, continuing his activities in pure research until very close to his death. In other respects, I never knew anyone quite like him.

Notes and references

(In these notes D. stands for P. A. M. Dirac.)

1 Parts of this paper were taken from A. Pais, *Inward Bound*, Oxford Univ. Press, Oxford and New York (1986).
2 T. Kuhn, interview with D., May 7, 1963, Niels Bohr Library, American Institute of Physics, New York.

3 Ref. 2, interview, April 1, 1962.

4 G. Gamow, *Thirty Years that Shook Physics*, p. 121, Doubleday, New York (1966).

5 For an account of D.'s Cambridge days see R. J. Eden & J. C. Polkinghorne, in *Aspects of Quantum Theory*, p. 1, eds. A. Salam & E. P. Wigner, Cambridge Univ. Press (1972).

6 D., in *History of Twentieth Century Physics*, p. 109, Academic Press, New York (1977).

7 D., Report KFKI-1977-62, *Hung. Ac. of Sc.*

8 D., *Proc. Roy. Soc.* **A109**, 642 (1925).

9 M. Born, *My Life*, p. 226, Scribner, New York (1978).

10 Ref. 6, p. 86.

11 Cf. D., *Proc. Camb. Phil. Soc.* **23**, 412 (1926).

12 D., *Proc. Roy. Soc.* **A112**, 661 (1926). For many more details about Dirac's early years and his contributions to quantum mechanics during 1925–6, see J. Mehra & H. Rechenberg, *The Historical Development of Quantum Theory*, Vol. 4, part 1, Springer, New York (1982); also L. Brown & H. Rechenberg, this volume.

13 W. Heisenberg, *Zeitschr. f. Phys.* **38**, 411 (1926).

14 W. Heisenberg, *Zeitschr. f. Phys.* **39**, 499 (1926).

15 E. Fermi, *Rend. Lincei* **3**, 145 (1926); *Zeitschr. f. Phys.* **36**, 902 (1926); repr. in *Enrico Fermi, Collected Works*, Vol. 1, pp. 181, 186, Univ. of Chicago Press (1962). In ref. 6, pp. 133, 134 Dirac has given a charming account of the time sequence of his and Fermi's contributions.

16 For the history of quantum statistics in the days of the old quantum theory see ref. 1, Chap. 13, section (d).

17 D., *Proc. Roy. Soc.* **A113**, 621 (1927).

18 Rigorous treatments lead to the theory of distributions, cf. I. Halperin & L. Schwartz, *Introduction to the Theory of Distributions*, Toronto Univ. Press (1952).

19 D., *Proc. Roy. Soc.* **A114**, 243 (1927).

20 D., *Proc. Roy. Soc.* **A114**, 710 (1927).

21 This so-called semi-classical procedure (discussed in detail by W. Pauli, *Handbuch der Physik*, Vol. 24/1, sections 15, 16, Springer, Berlin (1933)) allows for a good approximate but not rigorous treatment of induced processes; radiative corrections are not properly accounted for.

22 A. Einstein, *Phys. Zeitschr.* **18**, 121 (1917). See further A. Pais, *Subtle is the Lord*, Chap. 21, section (d), Oxford Univ. Press, New York (1982).

23 D. was aware[19] that he missed a factor two in this coefficient because he had not yet treated polarization properly.

24 Independently of Schroedinger, *Ann. der. Phys.* **81**, 109 (1926).

25 A far more detailed analysis of D.'s two founding papers on quantum electrodynamics has been given by R. Jost, in *Aspects of Quantum Theory*, p. 61, eds. A. Salam & E. P. Wigner, Cambridge Univ. Press (1972).

26 Ref. 20, p. 719.

27 D., *Proc. Roy. Soc.* **A111**, 405 (1926); *Proc. Camb. Phil. Soc.* **23**, 500 (1926).

28 A. Einstein, in *James Clerk Maxwell*, p. 66, Macmillan, New York (1931).

29 D., *Proc. Roy. Soc. Edinburgh* **59**, 122 (1939).

30 D., *Sci. Am.* **208**, 45 (May 1963).

31 O. Klein, *Zeitschr. f. Phys.* **37**, 895 (1926); E. Schroedinger, *Ann. der Phys.* **81**, 109 (1926); V. Fock, *Zeitschr. f. Phys.* **38**, 242 (1926); Th. de Donder & H. van den Dungen, *Comptes Rendues* **183**, 22 (1926); J. Kudar, *Ann. der Phys.* **81**, 632 (1926); W. Gordon, *Zeitschr. f. Phys.* **40**, 117 (1926).

32 W. Pauli, *Zeitschr. f. Phys.* **43**, 601 (1927).

33 He was looking for a four-dimensional generalization of $\sigma \cdot p$. Later he was to play briefly with wave equations for higher spin, D., *Proc. Roy. Soc.* **A155**, 447 (1936).

34 D., *Int. J. Theor. Phys.* **21**, 603 (1982).

35 D., *Proc. Roy. Soc.* **A117**, 610 (1928).

36 D., *Proc. Roy. Soc.* **A118**, 351 (1928).

37 Ref. 2, interview May 14, 1963.

38 It was later shown by Pauli and Weisskopf (*Helv. Phys. Acta* **7**, 709 (1934)) that the scalar wave equation is amenable to a treatment compatible with the transformation theory.

39 W. Heisenberg, interviewed by T. Kuhn, July 12, 1963, Niels Bohr Library, American Institute of Physics, New York.

40 D., *Phys. Zeitschr.* **29**, 561, 712 (1928).

41 W. Heisenberg, letter to W. Pauli, May 3, 1928, repr. in *Wolfgang Pauli, Scientific Correspondence*, Vol. 1, p. 443, Springer, New York (1979); referred to as PC below.

42 Ref. 40, p. 562, footnote 2.

43 W. Heisenberg, letter to W. Pauli, July 31, 1928, PC, Vol. 1, p. 466.

44 D., letter to O. Klein, July 24, 1928, copy in Niels Bohr Library.

45 S. F. Tuan, *Dirac and Heisenberg in Hawaii*, unpublished manuscript.

46 D., letter to A. Pais, October 21, 1982.

47 H. Weyl, *Zeitschr. f. Phys.* **56**, 330 (1929).

48 Other pertinent developments which had meanwhile taken place include the derivations of the Klein–Nishina formula for Compton scattering; and of the Klein paradox. See further ref. 1, Chap. 15, section (f).

49 D., *Proc. Roy. Soc.* **A126**, 360 (1929); also *Nature* **126**, 605 (1930).

50 D., ref. 6, p. 144.

51 See ref. 1, Chap. 14.

52 D., letter to N. Bohr, November 26, 1929, copy in Niels Bohr Library.

53 D., *Proc. Camb. Phil. Soc.* **26**, 361 (1930).

54 J. R. Oppenheimer, *Phys. Rev.* **35**, 562 (1930).

55 I. Tamm, *Zeitschr. f. Phys.* **62**, 545 (1930).

56 H. Weyl, *The Theory of Groups and Quantum Mechanics*, pp. 263–4 and preface, Dover, New York.

57 D., *Proc. Roy. Soc.* **A133**, 60 (1931).

58 C. D. Anderson, *Science* **76**, 238 (1932).

59 C. D. Anderson, *Phys. Rev.* **43**, 491 (1933).

60 W. Pauli, letter to D., May 1, 1933, PC, Vol. 2, p. 159.

61 W. Pauli, letter to W. Heisenberg, June 16, 1933, PC, Vol. 2, p. 169.

62 Cf. ref. 1, Chap. 16, section (d).

63 *The Theory of the Positron and Related Topics*, report of a seminar conducted by W. Pauli, notes by B. Hoffmann, Institute of Advanced Studies, Princeton (1935–6), mimeographed notes.

64 W. Heisenberg, letter to W. Pauli, PC, Vol. 2, p. 386.

65 D., letter to N. Bohr, August 10, 1933, copy in Niels Bohr Library.

66 The existence of vacuum polarization was also independently diagnosed by W. H. Furry & J. R. Oppenheimer, *Phys. Rev.* **45**, 245, 343 (1934).

67 D., in *Rapports du Septième Conseil de Physique*, p. 203, Gauthier-Villars, Paris (1934); cf. also D., *Proc. Camb. Phil. Soc.* **30**, 150 (1934).

68 A numerical error in the coefficient of that finite term was corrected by W. Heisenberg, *Zeitschr. f. Phys.* **90**, 209 (1934).

69 E. Uehling, *Phys. Rev.* **48**, 55 (1935).

70 Ref. 6, p. 140.

71 D., *Phys. Rev.* **74**, 817 (1948).

72 D., in dittoed notes of the Pocono conference, p. 72, unpublished.

73 D., in *New Pathways in Science*, Vol. 1, ed. A. Perlmutter, Plenum Press, New York (1976); see further E. Amaldi & N. Cabibbo, in *Aspects of Quantum Theory*, ref. 5, p. 183.

74 D., *Rev. Mod. Phys.* **21**, 392 (1949).

75 D., *Phys. Rev.* **73**, 1092 (1948); *Proceedings of the Second Canadian Mathematical Congress 1949*, p. 10, Univ. of Toronto Press, Toronto (1951).

76 D., *Can. J. Math.* **2**, 129 (1950), **3**, 1 (1951); *Proc. Roy. Soc.* **A246**, 326 (1958); *Proc. Roy. Irich Acad.* **A63**, 49 (1964).

77 See also F. Rohrlich, in *High Energy Physics*, p. 17, eds. B. Kursunoglu & A. Perlmutter, Plenum Press, New York (1985).

78 D., *Proc. Roy. Soc.* **A246**, 333 (1958); *Phys. Rev.* **114**, 924 (1959); also in *Recent Developments in General Relativity*, p. 191, Pergamon Press, London (1962). See further D., *Proc. Roy. Soc.* **A270**, 354 (1962); *Gen. Rel.*

and Grav. **5**, 741 (1974).

79 D., *Phys. Rev. Lett.* **2**, 368 (1959); *Proceedings of the Royaumont Conference 1959*, p. 385, Editions du CNRS, Paris (1962); *Phys. Bl.* **16**, 364 (1960).

80 D., *Ann. of Math.* **37**, 429 (1935).

81 D., *Ann. of Math.* **36**, 657 (1935).

82 D., in *Max Planck Festschrift 1958*, p. 339, Deutscher Verlag der Wissenschaften, Berlin (1958).

83 D., *General Theory of Relativity*, Wiley, New York (1975).

84 Ref. 6, p. 149.

85 D., *Nature* **139**, 323, 1001 (1937); also *ibid.* **192**, 441 (1961).

86 D., Report CTS-T.Phys. 69-1, Center for Theoretical Studies, Coral Gables, Florida (1969); *Comm. Pontif. Acad. of Sci.* **2**, No. 46 (1973); **3**, No. 6 (1975); *Proc. Roy. Soc.* **A338**, 446 (1974); *Nature* **254**, 273 (1975); in *Theories and Experiments in High Energy Physics*, p. 443, eds. B. Kursunoglu *et al.*, Plenum Press, New York (1975); *New Frontiers in High Energy Physics*, p. 1, eds. A. Perlmutter & L. Scott, Plenum Press, New York (1978); *Proc. Roy. Soc.* **A365**, 19 (1979).

87 See further F. J. Dyson, in *Aspects of Quantum Theory*, ref. 5, p. 213.

88 D., *Nature* **168**, 906 (1951); **169**, 146 (1952); *Physica* **19**, 888 (1953); *Sci. Monthly* **78**, 142 (1954).

89 D., *Proc. Roy. Soc.* **A136**, 453 (1932).

90 Cf. e.g. D., V. Fock & B. Podolsky, *Phys. Zeitschr. der Sowjetunion* **2**, 468 (1932).

91 D., *Nature* **137**, 298 (1936).

92 Ref. 1, Chap. 16, section (e).

93 Ref. 1, Chap. 18, section (a).

94 D., *Proc. Roy. Soc.* **A167**, 148 (1938); see also ref. 77.

95 D., *Ann. Inst. H. Poincaré* **9**, 13 (1939).

96 D., *Comm. Dublin Inst. Adv. Studies* **A1** (1943).

97 D., *Proc. Roy. Soc.* **A180**, 1 (1942).

98 D., *Rev. Mod. Phys.* **15**, 175 (1943).

99 D., *Comm. Dublin Inst. Adv. Studies* **A3** (1946); *Proceedings of the International Conference on Fundamental Particles and Low Temperatures, Cambridge, June 1946*, p. 10, Taylor and Francis, London (1946); *Proceedings of the 8th Solvay Conference 1948*, p. 282, ed. R. Stoops, Coudenberg, Brussels (1950).

100 R. P. Feynman, *Quantum Electrodynamics, the Strange Story of Light and Matter*, Princeton Univ. Press (1985).

101 D., *Proc. Roy. Soc.* **A209**, 251 (1951).

102 See also D., in *Deeper Pathways in High Energy Physics*, eds. B. Kursunoglu *et al.*, Plenum Press, New York (1977).

A. Pais

103 See further D., *Proc. Roy. Soc.* **A212**, 330 (1952); **223**, 438 (1954); also D., *Proc. Roy. Soc.* **A257**, 32 (1960); **268**, 57 (1962).
104 D., *Lectures on Quantum Field Theory*, Belfer School of Science, Yeshiva Univ., New York (1966).
105 Cf. D., *Nuov. Cim. Suppl.* **6**, 322 (1957); *Nature* **203**, 115 (1964); **204**, 771 (1964); *Phys. Rev.* **139B**, 684 (1965).
106 D., *Proc. Roy. Soc.* **A322**, 435 (1971); **328**, 1 (1972); and in *Fundamental Interactions in Physics and Astrophysics*, p. 354, ed. G. Iverson, Plenum Press, New York (1973).
107 D., in *Proceedings of Loyola University Symposium, New Orleans, 1984*.
108 Dirac was at the Institute for the academic years 1934–5, 1947–8, 1958–9, 1962–3; and for the fall term of 1946.
109 *Washington Post* and *Times Herald*, September 24, 1954.
110 E.g. *New York Times*, May 27, June 11, 1954; *New York Herald Tribune*, May 28, 1954; *The Times* (London), June 18, 1955; *The Financial Times* (London), August 6, 1954.
111 *New York Times*, June 3, 1954.
112 D., *Proc. Camb. Phil. Soc.* **35**, 416 (1939).
113 Ref. 74, p. 393.
114 D., *The Development of Quantum Theory*, Gordon and Breach, New York (1971).
115 D., *Eureka* No. 32, 2–4 (October 1969).

116

12

Paul Dirac and Werner Heisenberg – a partnership in science

Laurie M. Brown*
Northwestern University, Evanston, Illinois

and

Helmut Rechenberg
Max-Planck Institut für Physik und Astrophysik, West Germany

During an interview carried out in 1963, Carl Friedrich von Weizsäcker, one of Heisenberg's earliest students, was asked what the latter thought about the possibility of electrons in the nucleus (before the discovery of the neutron). After noting that the problem was related to that of relativistic motion of the electrons, von Weizsäcker continued:

I remember that Heisenberg said, 'Well now, the problem of relativity and quantum theory and the electron has been solved by a young Englishman by the name of Dirac, and he is so clever it is not . . . *es lohnt sich nicht, mit dem um die Wette zu arbeiten.*' And I felt that Heisenberg was just really unhappy to see that there was a man who was able to solve a problem which he hadn't been able to do himself (1).

On the other hand, Dirac introduced Heisenberg to an audience in 1968 with *this* comment on Heisenberg's 1925 invention of matrix mechanics:

I have the best of reasons for being an admirer of Werner Heisenberg. He and I were young research students at the same time, about the same age, working on the same problem. Heisenberg succeeded where I failed. There was a large mass

* L.M.B. is grateful to the Max-Planck Institut für Physik und Astrophysik, Werner-Heisenberg-Institut für Physik, for hospitality and summer support. He also wishes to thank the Program in History and Philosophy of Science of the National Science Foundation, U.S.A., for a grant that assisted this work.

of spectroscopic data accumulated at that time and Heisenberg found the proper way of handling it. In doing so he started the golden age in theoretical physics, and for a few years after that it was easy for any second rate student to do first rate work (2).

Dirac might have added that he was, himself (already in the fall of 1925), among the first to help in creating that golden age, and that his own contributions paved the way to unexpected extensions of the theory. In 1933, when Heisenberg received the 1932 Nobel Prize for Physics and Dirac shared the 1933 prize with Erwin Schrödinger, Heisenberg wrote to Niels Bohr that he wished the full prize had been given to Dirac (and to Schrödinger as well). Indeed, after quantum mechanics, there was no development which Heisenberg rated more highly than the relativistic electron equation of Dirac. About thirty years later, Heisenberg's fundamental field equation, intended to describe all elementary particles and their interactions, was closely modelled after Dirac's electron equation of 1928, and a crucial rôle in the new theory was ascribed to the indefinite metric, which had been proposed by Dirac in his Bakerian Lecture of 1941.

For nearly fifty years, they worked on the forefront of theoretical, atomic, molecular, and elementary particle physics, their programs often overlapping. The two met first at Cambridge in July 1925 and became close personal friends in the fall of 1926 at Bohr's institute in Copenhagen. Afterwards, they saw each other frequently, either at home or when travelling, and their lifelong friendship was a unique one for both of them. On the present occasion, we want to recall some of their fruitful scientific and personal exchanges during three periods: 1925 to 1927 (discovery and completion of quantum mechanics), 1928 to 1932 (relativistic electron equation and quantum electrodynamics), and 1935 to 1976 (quantum fields and elementary particle theory).

1 Discovery and completion of quantum mechanics (1925–27)

Dirac's introduction, quoted above, referred to this situation: Heisenberg, who had been a student of Arnold Sommerfeld at Munich beginning in the fall of 1920, had then collaborated with Max Born in

The Nobel Prize Winners (Physics 1932 and 1933), Stockholm, December 1933. From left to right: Mrs A. Heisenberg (mother), Mrs H. Dirac (mother), Mrs A. Schrödinger (wife), Dirac, Heisenberg, Schrödinger.

Göttingen and Niels Bohr in Copenhagen, trying to solve the difficulties in the Bohr–Sommerfeld quantum theory of atomic structure. At the same time, the equally youthful Dirac, who had graduated in electrical engineering in 1921 and applied mathematics in 1923 at Bristol and had then become a research student of Ralph Fowler at Cambridge, was working on the same problems. Heisenberg, older by only eight months, developed a new theoretical scheme in July 1925, which Dirac recognized as the 'key to the whole mystery' (3).

1.1 Dirac's reaction to Heisenberg's pioneering paper

Heisenberg's paper, 'Über die quantentheoretische Umdeutung kinematischer und mechanischer Beziehungen,' finished on 9 July 1925, proposed to describe dynamical variables (such as the coordinates and momenta of electrons) by symbols depending on two quantum numbers, and obeying a multiplication law that was not commutative (4). He applied this scheme to integrate the equations of motion of the anharmonic oscillator and of the rotator and achieved results in agreement with some spectroscopic data (namely, on molecular spectra and the multiplet structure of atoms). During a visit to Cambridge to present a talk at the so-called Kapitza Club on 28 July 1925, Heisenberg mentioned his new results to Fowler, who asked him to send a copy of the paper as soon as it was available. Heisenberg received the proofsheets of his article by the middle of August and mailed the second copy to Fowler, who then passed them on to Dirac to study.

Dirac set the paper aside for a week or so, but he soon realized how important it was and tried to relate Heisenberg's new dynamical quantities to the Hamilton–Jacobi action-angle variables. After weeks of intensive research, he got his decisive idea: 'During a long walk on a Sunday, it occurred to me that the commutator (i.e., the difference $xy - yx$, where x and y denote two of Heisenberg's non-commuting dynamical variables) might be the analogue of the Poisson bracket' (5). Explicitly, if p_r and q_r are a complete set of canonically conjugate variables used in the classical description of an atomic system – the p_r denoting momentum and the q_r position coordinates – and x and y are functions of them, then the following association exists between them and Heisenberg's x and y (placed on the left-hand side):

$$xy - yx \leftrightarrow \frac{ih}{2\pi} \sum_r \left(\frac{\partial x}{\partial q_r} \frac{\partial y}{\partial p_r} - \frac{\partial y}{\partial q_r} \frac{\partial x}{\partial p_r} \right).$$

From the end of September to early November 1925, Dirac worked out a theory based on this correspondence, described in his paper, 'The Fundamental Equations of Quantum Mechanics' (6). It should be noted that, about six weeks earlier, Max Born and Pascual Jordan in Göttingen had submitted to *Zeitschrift für Physik* their paper, 'Zur Quantenmechanik', in which they formulated Heisenberg's scheme in matrix language (7). Then, Born, Heisenberg, and Jordan further developed matrix mechanics at about the same time that Dirac worked out his theory (8). When Dirac finished his manuscript, he sent Heisenberg a copy, who acknowledged it immediately in a letter dated 20 November 1925.

Many years later, Dirac said, 'That was really ... a very kind letter, because he did not want me to be disturbed by the fact that there were other people simultaneously working on these problems who had anticipated my results to some extent, and he was also full of praise for my own contributions' ((9), p. 126). Heisenberg had especially appreciated Dirac's connection between the quantization conditions and Poisson's brackets, and the general definition of differential quotients in quantum mechanics. He also declared that Dirac's paper was better written than 'our attempts here.'[1] That Heisenberg was not merely flattering Dirac can be seen from a letter he wrote on the same day to Niels Bohr:

Today, I received a work sent me by Dirac, in which he has done the mathematical part of the new quantum mechanics on the basis of my work (independently of Born and Jordan) ... in its style in writing, some of it pleases me better than that of Born and Jordan.[2]

Dirac went ahead to show that his q-number formulation of quantum mechanics was superior to matrix mechanics by solving the problem of the hydrogen spectrum (10), and especially by generalizing his scheme to apply to many-electron atoms (11) and certain relativistic problems, such as Compton scattering (12). The second paper, containing the application to the hydrogen atom, brought forth this response from Heisenberg:

I congratulate you. I was quite thrilled as I read the paper. Your separation of the problem into two parts – calculation with 'q-numbers,' on the one hand, and physical interpretation of the 'q-numbers,' on the other hand – seems to correspond, in my opinion, completely to the nature of the mathematical problem. With your treatment of the hydrogen atom, there seems to me a small step towards the calculation of the transition probabilities (i.e., the intensity of spectral lines).'[3]

He then went on to ask Dirac's opinion about Schrödinger's work on the hydrogen atom:

A few weeks ago an article by Schrödinger appeared ... whose contents to my mind should be closely connected with quantum mechanics. Have you considered how far Schrödinger's treatment of the hydrogen atom is connected with the quantum mechanical one? This mathematical problem interests me especially because I believe that one can win from it a great deal for the physical significance of the theory.[4]

1.2 Wave mechanics, transformation theory, and uncertainty relations

At first, Dirac hesitated to accept Schrödinger's wave mechanics; he was happier with his own scheme, and he criticized Schrödinger in his response to Heisenberg, who wrote again on 26 May 1926 (this time in English): 'I quite agree with your criticism of Schrödinger's paper with regard to a wave theory of matter. This theory must be inconsistent, just like the wave theory of light.' But he added, 'I see the real progress made by Schrödinger's theory in this: that the same mathematical equation can be interpreted as point mechanics in a non-classical kinematics *and* as wave theory according to Schrödinger. I always hope that the solution of the paradoxes in quantum theory later could be found in this way.' In the same letter, Heisenberg discussed the formal connection between Schrödinger's wave mechanics and the Göttingen–Cambridge quantum mechanics, showing how the physical quantities in the two schemes were related.[5]

On Heisenberg's urging, Dirac returned to the study of Schrödinger's paper (15), analyzed it, and derived further interesting results; then, in August 1926, he submitted a paper in which he showed that the original Born–Jordan formulation of the energy eigenvalue problem for a matrix Hamiltonian corresponded to using a Schrödinger equation with a special choice of the eigenfunctions (16). He found also that the wave function of a two particle system could be written in either a symmetric or an antisymmetric way, i.e.,

$$\psi_{mn}(1,2) = \psi_m(1)\psi_n(2) \pm \psi_m(2)\psi_n(1).$$

For the case that the minus sign held in this equation, he determined the statistical behavior of the particles, obtaining for their distribution as a function of temperature:

$$N_s = \frac{A_s}{\exp(a + E_s/kT) + 1}.$$

(N_s is the number of particles having the energy E_s, k being Boltzmann's constant, T the absolute temperature, and a another constant.) This distribution is different from that of Bose–Einstein statistics, and applies to particles that obey Pauli's exclusion principle – as Enrico Fermi had, in fact, shown earlier (17).[6]

In the middle of September 1926, Dirac arrived in Copenhagen to spend five months at Bohr's institute. There he met Heisenberg, who, in May, had replaced Hendrik Kramers as assistant to Bohr and lecturer at the University of Copenhagen. The fruitful cooperation between the two young physicists would continue.

Heisenberg had spent the first half of 1926 continuing his pioneering investigations into quantum mechanics. In March, he had finished a paper with Jordan, dealing with the long-standing problem of the anomalous Zeeman effect. Using the action-angle formalism of Dirac (which had also been independently developed by Gregor Wentzel on the basis of matrix mechanics (18)) and including the spin, they had derived the successful formula that had been obtained earlier semi-empirically (19). They also obtained the relativistic fine-structure formula for hydrogen-like atoms. While in Copenhagen, Heisenberg had treated the problem of resonance in quantum mechanical systems and had introduced the concept of exchange energy (20). Then, in June and July, he made a successful attack on the energy levels of the helium atom, an old acquaintance of his, and also of Dirac (21).

In evaluating certain matrix elements required for the helium problem, Heisenberg had used wave-mechanical methods, but, after the middle of 1926, he became increasingly critical of Schrödinger's physical interpretation, for the latter claimed to have shown that he had eliminated discontinuous quantum jumps from the theory. Heisenberg, on the other hand, now rather emphasized the discontinuous aspects of quantum theory, as shown in his report, given on 2 September 1926 at the Düsseldorf Naturforscherversammlung (22). Upon his return from that meeting to Copenhagen, he worked on two papers in which he tried to make the consequences of quantum mechanics more precise, while avoiding altogether any reference to wave-mechanical methods (23), (24).

In the paper on fluctuation phenomena in quantum mechanics, submitted in November, Heisenberg demonstrated that, in the quantum mechanical interaction of two identical atomic systems, although there is an analogy to the classical resonance phenomenon, in which energy is transmitted from one vibrating system to the other at a slower beat frequency, one cannot speak of the energy as being transmitted continuously as a function of time, as Schrödinger had claimed; he showed, instead, that, for any given state of the total system, only time averages have physical significance:

The calculations carried through here appear to me to be an argument that a continuous interpretation of the quantum mechanical formalism, thus also that of the de Broglie–Schrödinger waves, would not correspond to the essential meaning of the well-known formal connections.[7]

The result that Heisenberg stated here so emphatically cast light on the relation between the theory and its experimental implementation, a problem to which Dirac now began to pay increasing attention. Having learned from Heisenberg about this result before its publication, he claimed it to be 'capable of wide extensions,' for, he said:

It can be applied to any dynamical system, not necessarily composed of two parts in resonance with one another, and to any dynamical variable, not necessarily one that can take only quantized values... It thus appears that certain questions that one can ask about the classical theory can be given unambiguous answers on the quantum theory as well as on the classical theory.

This was contained in the introduction to his next paper, 'The Physical Interpretation of the Quantum Dynamics,' sent for publication in early December 1926 (see (25), p. 622). In it, he established a systematic transformation theory, with which he showed that there existed a one-to-one relation between the Göttingen–Cambridge quantum mechanics and Schrödinger's wave mechanics, but without the latter's physical interpretation.[8]

While Dirac was working out those results in Copenhagen, Jordan was developing another transformation scheme in Göttingen; he sent his manuscript to Heisenberg, who showed it to Dirac. After having studied it, Dirac wrote to Jordan, 'As far as I can see it is equivalent to my work in all essential parts. The way of obtaining the results may be rather different though.' And after explaining this point, he concluded, 'I hope you do not mind the fact that I have obtained the same results as you, at (I believe) the same time.'[9]

Whatever the merits of the two authors' approaches, Heisenberg made extensive use of their investigations when he derived his uncertainty relation,

$$\Delta p \cdot \Delta q \geqslant h/4\pi,$$

which states that the precision of measurement of position and momentum is limited by the nonzero value of Planck's quantum of action h. And he wrote in his famous paper, 'Über den anschaulichen Inhalt der quantentheoretischen Kinematik und Mechanik': 'This uncertainty is the real basis for the occurrence of statistical relations in quantum mechanics,' adding, 'Their mathematical formulation can be achieved by means of the Dirac–Jordan theory. Departing from the foundations so achieved, it will be shown how macroscopic processes can be understood from the quantum mechanical point of view.'[10]

Thus, in less than two years – Heisenberg submitted his paper on the *anschauliche* interpretation in March 1927 – Dirac's and Heisenberg's combined efforts succeeded in establishing both the mathematical theory and the physical interpretation of quantum mechanics. This theory would be presented and intensively discussed at the Fifth Solvay Conference, held in Brussels in October 1927.[11]

2 Relativistic quantum theory

'The new quantum mechanics, when applied to the problem of the structure of the atom with point-charge electrons, does not give results in agreement with experiment.' So begins Dirac's landmark paper, 'The Quantum Theory of the Electron' (received 2 January 1928), in which he

Laurie M. Brown and Helmut Rechenberg

proposes his relativistic equation for the spinning electron and shows that it produces, at least to first order of accuracy, the relativistic correction to the hydrogen spectrum as well as the 'duplexity' phenomena shown in spectra; that is, the doubling of the number of states, corresponding to the electron having one-half quantum of angular momentum, or spin. In Part II of that work (received one month later), Dirac uses his new equation to calculate successfully the relative intensity of the spectral lines of the anomalous Zeeman effect (29). Shortly thereafter, both Charles Galton Darwin (30) and Walter Gordon (31) showed that the relativistic fine structure formula for the hydrogen energy levels follows exactly from the Dirac equation. More remarkable still, when the theory was found to have 'objectionable' negative energy states, Dirac's new physical interpretation of these states in his so-called 'theory of holes' predicted a new elementary particle, the positive electron or positron (32), whose presence in the cosmic rays was confirmed in August 1932 by Carl Anderson (33). Soon after that, Patrick M. S. Blackett and Giuseppe P. S. Occhialini found Dirac's predicted electron–positron pairs in their counter-triggered cloud chamber (34).[12]

2.1 Heisenberg's response to the electron equation

One might easily argue that Dirac's relativistic electron equation, his hole theory, and his pioneering work on quantum electrodynamics (35) were the principal theoretical advances during the years 1927–32.[13] But, in 1969, when Heisenberg looked back upon that period, he wrote:

To those of us who participated in the development of atomic theory, the five years following the Solvay Congress in Brussels (in 1927) looked so wonderful that we often spoke of them as the golden age of atomic physics. The great obstacles that had occupied all our efforts in the preceding years had been cleared out of the way; the gate to that entirely new field – the quantum mechanics of the atomic shell – stood wide-open, and fresh fruits seemed ready for the plucking. Where purely empirical rules or vague concepts had had to serve as substitutes for real understanding – for instance, of ferromagnetic phenomena and of chemical bonds in the physics of solids – the new methods brought absolute clarity. Moreover, it seemed very much as if the new physics was in many respects greatly superior to the old even on the philosophical plane; that, in ways that had to be investigated more closely, it was much broader and richer (36).

126

Note that Heisenberg does not even mention here the relativistic quantum theory!

In the AHQP interview with von Weizsäcker (which we quoted earlier), Thomas S. Kuhn asked, 'Did you get the impression that Heisenberg himself had been working on the relativistic problem? In his published papers from this period there's just nothing at all.' 'No, nothing at all,' von Weizsäcker agreed, then added, '... perhaps one may say that Heisenberg was surprised to see that it was possible to invent a relativistic equation which seemed to be consistent. I mean, he might have expected that relativity and quantum theory were so different that it would need far greater steps than the steps done by Dirac, not just the step to invent a relativistic equation' (1). From the foregoing, one might surmise (incorrectly) that Heisenberg was not involved with the problem of combining relativity with quantum mechanics. That was not the case. He regarded that problem as one of fundamental importance, but, until the puzzle of the negative energy states in Dirac's theory could be solved, Heisenberg remained sceptical – an attitude prevalent among the adherents of the Copenhagen school. Negative total energies have no meaning and are not acceptable in the theory of relativity; but even more important, according to Heisenberg, electrons in positive energy states would spontaneously drop into negative energy states (emitting a photon to carry off the lost energy and momentum). That would leave no electrons to observe. Dirac's eventual suggestion, that this could be avoided by all the negative energy states being filled, was not regarded favorably. Thus, in March 1928, Heisenberg wrote to Bohr from Munich: 'I have marvelled greatly at Dirac's works; but I find it very disturbing that such an apparently so complete theory as that of Dirac (of the relativistic electron) shows such terrible defects as transitions from positive to negative energy.'[14]

Three months later, he wrote to Jordan from Leipzig, where he had accepted his first professorship:

Dirac has lectured here only on his current theory, giving as a pretty foundation for it that the differential equations must be linear in $\partial/\partial x_\mu$. He has not been able to solve the well-known difficulties; I have discussed them thoroughly with Dirac; I personally tend to believe that one must try somehow to bring in the asymmetry in $+e$ and $-e$ (m and M). A strong argument for that is also that supposedly the ratio m/M is connected with the fine structure constant. Dirac is, however, of another opinion and still hopes to obtain a contradiction-free theory for a single electron.[15]

He wrote to Bohr, in much the same spirit, during the following month, that '... upon more careful reflection the difficulties are much more serious than they appeared to me in the beginning.' On the one hand, the radiative transitions to negative energy states seemed to be much more frequent than any other spontaneous transitions; on the other hand, if one looked at the scattering of light by electrons, one could see that those processes could not be simply ignored, for '... the ominous processes are still an integral constituent of Dirac theory.' Heisenberg concluded with the revealing remark: 'Thus I find the present situation quite absurd and on that account, almost out of despair, I have taken up another field, that of ferromagnetism.'[16]

Heisenberg's interest in making a quantum mechanical theory of magnetism went back to the summer of 1926, when he realized that the exchange interaction that he had discovered in treating the states of the helium atom could provide the strong spin-dependent force needed to align the spins (hence, also the magnetic moments) of the electrons. A little later, when Heisenberg was in Copenhagen together with Dirac, and wrote to Pauli about Dirac's first attempt to unite relativity and quantum theory,[17] he went on to say:

I myself have thought a bit about the theory of ferromagnetism, about conductivity and similar dirty things. The idea is this: In order to use Langevin's theory of ferromagnetism, one must assume a large coupling force between the spinning electrons (*only these* are turning around (in a ferromagnet)). This force should, as for helium, be indirectly supplied by resonance. I believe that one can always prove: parallel position of the spin vectors always gives the *smallest* energy ... I have the feeling that this in principle could be sufficient to provide an explanation of ferromagnetism.[18]

In October 1927, after one and a half years in Copenhagen, Heisenberg went to Leipzig to occupy the chair of theoretical physics at the university. He and Peter Debye, who had just assumed the chair of experimental physics, made Leipzig an important center for research and instruction in the new physics. Pauli, who had been a professor in Hamburg since November 1926, was called to a chair for theoretical physics – actually the chair vacated by Debye – in Zürich in April 1928. Earlier, Pauli and Sommerfeld had originated the quantum theory of metallic conduction, and, beginning in 1928, the schools of Heisenberg and Pauli, in collaboration or separately over the next five years produced great advances in the application of quantum mechanics to solids, a field that would eventually be called solid state theory. Some of Heisenberg's students working on the properties of the solid state during that period were: Felix Bloch, Rudolf Peierls, and Edward Teller. (His other students included von Weizsäcker, Hans Euler, Leon Rosenfeld, and Gian Carlo Wick.)

Meanwhile, Dirac, who had come to Cambridge in 1923 as a graduate student, completed his doctoral degree in May 1926 and became Fellow of St. John's College of that university in 1927. His duties included some lecturing, but mostly on his own recent research. Dirac was typically a loner, throughout his life only rarely collaborating on scientific work with others, and even after he was appointed to the Lucasian Chair of Mathematics (a position that had been held by Isaac Newton), he accepted few research students.[19] Thus, while Heisenberg and Pauli felt obliged to spend at least part of their efforts on applications of quantum mechanics that would provide good problems for their research students, Dirac was able to concentrate on more fundamental questions.[20] In noting this point, however, one should keep in mind that both Pauli and Heisenberg had been trained in Munich in the school of Arnold Sommerfeld, a great teacher who was a master of applied mathematical physics, and they never forgot the lessons they learned from him. Furthermore, as we shall discuss, they did carry out fundamental research on the relativistic quantum theory of fields.

2.2 Relativistic quantum field theory

Heisenberg and Pauli's reformulation of quantum electrodynamics was begun before the time that Dirac was inventing his relativistic electron theory.[21] Paul wrote to Dirac from Hamburg on 17 February 1928, saying that he and Heisenberg were occupied with the problem of giving a relativistically invariant formulation to the quantum theory of electromagnetic interaction. (In the same letter, he said that he was sending Dirac the manuscript of a paper by Pascual Jordan and Eugene Wigner, in which they quantized the field representing fermions, i.e., they introduced the idea of second quantization (41).) Pauli briefly explained the idea behind the Heisenberg–Pauli work: to let the

129

electromagnetic field quantities, as well as the matter fields, be *q*-numbers obeying specified commutation relations, while satisfying the classical field equations as supplementary conditions. In their case, the matter fields were to obey the relativistic Schrödinger (also known as the Klein–Gordon) equation.[22] While they had not arrived at a final result, Pauli felt that they had made good progress ('ein gutes Stück vorwärts gekommen'). He urged Dirac to tell him the relationship between Dirac's new electron theory and his earlier version of quantum electrodynamics, and, again, he brought up the problem of spontaneous transitions to negative energy states (which he regarded as states of positive charge).[23]

On 19 March 1929 – Heisenberg being at the time in America – Heisenberg and Pauli submitted their paper, 'Zur Quantendynamik der Wellenfelder,' a major work on the Hamiltonian theory of relativistic fields, and a pioneering achievement not only in its quantum aspects but also in classical field theory. While they still had the expected difficulty in handling the Dirac electron field, they nevertheless tried to treat the effects of retarded forces as a separate problem; i.e., to treat fields transmitted with the velocity of light, without specializing (as Dirac had done) to the radiation field alone. As they noted in the introduction to their paper, they could not expect to be able to write down and solve the many-body equations of motion, since that could not be done even in classical mechanics. The correspondence principle was assumed in the sense that the quantum electromagnetic field should, under suitable conditions (large occupation numbers), approach the Maxwell field, while the electron field should be that satisfying Dirac's equation. In addition to finding an infinite charge density of the vacuum, if Dirac's 'holes' were filled, they encountered the problem of the 'infinities' that would plague the quantum field theory until the renormalization program of the 1940s. Specifically, Heisenberg and Pauli recognized the problems of infinite zero-point energy and infinite electron self-mass (43).[24]

Because of these difficulties, Heisenberg was less than enthusiastic about his work with Pauli and wrote to Bohr on 1 March 1929 from the ship that was taking him to America:

I am particularly eager to know what you will say about the relation between quantum theory and relativity theory. Incidental to the work with Pauli I have also pondered much on those questions of principle, but without much success; the work by Pauli and myself is, to begin with, at best a formal advance, perhaps not even that, because the Dirac difficulties naturally remain unsolved ... Whether I get to work on papers at all in America seems unclear to me; to my joy I heard that Dirac will also be over there (in America) in the next months, so that I will be able to discuss theory a bit.[25]

As we shall describe in the next section, Dirac and Heisenberg did meet in America, and visited the Far East together before returning to Europe. Back again in Leipzig, Heisenberg wrote to Bohr in December 1929:

Pauli's and my electrodynamics has further quite remarkably simplified itself; the earlier extra terms have all become entirely superfluous, and, aside from some formal ugliness (Unschönheiten), it seems to me that this interaction problem is now in order; however, it still remains a very gray theory, as long as the Dirac difficulty is unsolved. Dirac has already written a new work on the $\pm e$ business; you probably also know that. I am, however, really skeptical, because the proton mass comes out equal to the electron mass, so far as I can see.[26]

In order to appreciate the quandary that resulted from the 'ominous' negative energy states of Dirac's electron theory, it is necessary to appreciate the power of the theory. In the first place, it was relativistically covariant and gave exactly the Sommerfeld relativistic fine-structure formula, which appeared to fit perfectly the hydrogen spectrum. It gave correctly the spin angular momentum and the magnetic moment of the electron. These properties were, in a sense, built into the theory. But above all, perhaps, the most convincing argument for the theory's validity was given by the calculation by Oskar Klein and Yoshio Nishina, in Copenhagen, of the scattering of X-rays and gamma rays from free electrons (44). Their formulas gave good agreement with the intensity, the angular distribution, and the polarization of the scattered radiation up to the highest available known gamma ray energy – with one important exception.[27]

The expression was a rather large excess scattering (up to 40 percent), found when the most energetic known gamma ray ('Th C') scattered in the elements of highest atomic number (e.g., lead). The anomalous effect was found almost simultaneously by several experimental groups in America and Europe, who submitted their results for publication during May 1930. It became known as the Meitner–Hupfeld effect, and, for several years afterwards, it was thought to be a nuclear structure effect arising from electrons that were thought to be present in the nucleus. Eventually, it was recognized to be the combined effect of electron–positron pair production, subsequent annihilation of the positron, and electron and positron *Bremsstrahlung*. For our present purposes, its main significance is that it tells us that the predictions of Dirac's electron theory were accepted so completely that any deviation from them (even at the extreme limits of known high energy and high nuclear mass) was considered a major anomaly that heralded a new physical phenomenon.[28]

3 Dirac and Heisenberg as travelling companions

Dirac and Heisenberg each visited the United States for the first time in the spring of 1929, Dirac to spend a term as visiting professor at the University of Wisconsin in Madison, Heisenberg to lecture on quantum mechanics at the University of Chicago. They met in Madison and agreed that they would return to Europe via the Pacific. They did so, sharing passage on a Japanese ship from San Francisco to Yokohama, and stopping *en route* in Hawaii. After leaving Japan, however, they travelled separately – Heisenberg taking a southern route via Hong Kong and India, while Dirac took a ship to Vladivostock and then crossed Siberia by rail. Most of their trip was devoted to sightseeing, but they gave a series of lectures in Japan which had an important influence on the physics that was done there afterwards.

In an article of reminiscence about Dirac written in 1972, J. H. van Vleck told about the trip: 'Dirac decided that when he got as far west as Wisconsin he might as well go around the world, and after leaving Madison he made at least the first part of this trip along with Heisenberg. The latter was lecturing at the University of Chicago in the spring of 1929. While Dirac was in Madison, Heisenberg came to Madison for a brief trip to give a colloquium. Whether they arranged their voyage then, or before Dirac left England, I don't know.'[29] In fact, the idea of a joint visit to the Far East had been proposed to Dirac by Heisenberg more than a year earlier in the following letter, which we quote in full:

It is very probable that I will go to Chicago from April 1929 to Sept. 1929; I have

132

Dirac (left) and Heisenberg. Chicago, summer 1929.

decided, not to go in this year; the journey for next year is not quite certain yet. Of course I would be extremely glad, if we could work together those 6 months in Chicago and bring European life into the American hurry. If you go to Chicago, I will certainly go. Perhaps we could have some pleasure from seeing beautiful parts of the country, f.e. from seeing California, which I probably would visit in July or June. Or we could go back to Europe via Japan–India or China etc. But of course you ought to do, what you like best. – I admire your last work about the spin in the highest degree. I have especially still for questions: do you get the Sommerfeld-formula in *all* approximations? Then: what are the currents in your theory of the electron? – I am writing a paper together with Pauli. We tried to change the Schrödinger theory of the Broglie-waves + Maxwell-waves ('Energie–Impuls–tensor' etc.) into a theory of quantized waves (Bose or Fermi statistics does not make much difference); in so far the theory is generalization of Klein–Jordans work as well as of Pauli–Jordans last paper (relativistic invariant Vertauschungs-relations between the F_{ij}). On(e) gets new terms in the Energy–Impuls–tensor, which compensate the interaction of the electron with itself, like in Klein–Jordans work. I think, one gets a definite idea, how retarded potentials etc. have to be treated in quantum mechanics. Of course the theory has to be extended to the spin in your way. – You will get proofs of the paper, as soon as possible.[30]

Three days after Heisenberg wrote this letter, on 16 February 1928, Arthur Compton sent a telegram to Dirac at St. John's College, reading, 'Can you come next fall and winter Heisenberg comes spring and summer.' Apparently, Dirac did not accept Compton's invitation, but, instead, agreed to visit Madison during the following spring. On 23 January 1929, Heisenberg again wrote to Dirac: 'I will be in Cambridge (Massachusetts) in March and in Chicago from the beginning of April till the end of August. I am very glad to hear that you are going to America too, I think we will meet there several times.'[31] And from Cambridge he wrote (the letter is undated, but it must have been later in March, where Dirac was in Madison):

Next Friday I will arrive in Chicago; since we then are not more than 200 miles apart from another, I would like to establish the connection between us. Could you be kind and write to me, w(h)ether may visit you some weekend, say April 6th or 13th. Then I would be especially interested to hear your opinion about Weyl's work. Weyl thinks to have the solution of the $\pm e$ difficulty ... Shortly after your last letter to Leipzig I saw Eddington's paper on the charge e. I cannot find anything reasonable in it, but perhaps you understand the deeper reasons and can tell me about them. Sometimes I am entirely wrong in judging other papers. Before I left Germany, Pauli and I finished a paper about the relativistic formulation of the many-body problems; I would very much like to show it to you and to hear your opinion.[32]

In writing to Bohr from the *S.S. Washington* on his overseas trip, Heisenberg had questioned whether he would be able to produce any new works there.[25] As he had just completed his very important and extensive work with Pauli on the quantum theory of fields, and, before that, had proposed his new theory of ferromagnetism, it might have been a good occasion for him to rest and seek fresh inspiration. In the event, however, the lectures that he delivered at the University of Chicago in the spring of 1929 were sophisticated physical, mathematical, and even philosophical discussions of the fundamental aspects of quantum theory, including: critiques of the physical concepts of wave and particle (in the light of the uncertainty principle), the important underlying experiments, the statistical interpretation of quantum mechanics, complementarity, relativistic quantum mechanics and field theory. With the addition of an eighty page appendix on 'The Mathematical Apparatus of the Quantum Theory,' the lectures were published in English as Heisenberg's first book. It is a classic work, although Heisenberg's preface modestly disclaims:

On the whole the book contains nothing that is not to be found in previous publications, particularly in the investigations of Bohr. The purpose of the book seems to me to be fulfilled if it contributes somewhat to the diffusion of that '*Kopenhagener Geist der Quantentheorie*,' if I may so express myself, which has directed the entire development of modern atomic physics (45).

While Heisenberg was lecturing at Chicago, Dirac was sharpening the presentation of his own approach to quantum mechanics and laying down the foundation for *his* first book, probably the most famous and influential of all the books that have ever been written on the subject. Its first edition appeared in 1930, with subsequent editions in 1935, 1947, and 1958 (46). Concerning his term in Madison, van Vleck wrote: 'Dirac gave a well-organized course of lectures, almost a formal course, on mainly the transformation theory of quantum mechanics, which he had evolved about a year and a half earlier. It was far more abstract than the usual American courses in physics of that era.' The visit was important to van Vleck's work, as he related:

Dirac either lectured on, or told me about his now celebrated vector model (47) for handling permutation degeneracy. In particular he indicated how it could be used to obtain Heisenberg's formulas for ferromagnetism, something not mentioned in his publications. I was greatly influenced by Dirac's procedure, and subsequently capitalized on it heavily, applying it not only to magnetism but also to complex spectra and chemical bonding... All I did was apply it. Dirac had the original and essential idea.[33]

The residents of Madison, other than the physicists at the University, learned about Dirac from a local newspaper interview:

I been hearing about a fellow they have up at the U. This spring – a mathematical physicist, or something, they call him – who is pushing Sir Isaac Newton, Einstein and all the others off the front page... His name is Dirac and he is an Englishman... So the other afternoon I knocks at the door of Dr. Dirac's office in Sterling Hall and a pleasant voice says 'Come in.' And I want to say here and now that this sentence 'come in' was about the longest one emitted by the doctor during our interview.

I found the doctor a tall youngish-looking man, and the minute I seen the twinkle in his eye I knew I was going to like him ... he did not seem to be at all busy. Why if I went to interview an American scientist of his class, ... he would blow in carrying a big briefcase, and while he talked he would be pulling lecture notes, proofs, reprints, books, manuscripts, or what have you, out of his bag. Dirac is different. He seems to have all the time there is looking out the window... 'Professor,' says I, 'I notice you have quite a few letters in front of your last name. Do they stand for anything in particular?'

'No,' says he (etc.).[34]

For the return trip of Dirac and himself at the end of their American visits, Heisenberg made the travel arrangements as far as Japan; in May, he wrote Dirac that he would make the cabin reservations for both of them.[35] In June, he wrote, reporting that he was unable to get a cabin on the *President Taft* of the Dollar Line, as it was already fully booked. Instead, he had reserved passage on the N.Y.K. (Japanese) Line. After giving a number of details, such as the ship's displacement and gross tonnage, he said, 'I don't know, whether you mind, to sail on a Japanese steamer; but I did not know what to do else.'[36]

Receiving no reply from Dirac, he wrote the following on 5 July:

A few weeks ago I wrote a letter to you about our trip to Japan; apparently you did not get it. The content was: I was *not* able to get a cabin on *President Taft*, she was crowded. So I made the reservations on a Japanese steamer, leaving San

Francisco Aug. 14th. Our cabin is an outside cabin 1st class; we will arrive in Yokohama Aug. 30th. The price of the ticket is the same as on *President Taft*, about 300 $. The size of the steamer is about the same as that of *President Taft*. If you should not like to take the Japanese steamer, please write to me at once, if possible. – My lectures in Japan will probably be (– thanks to your letter!): 1) Principle of uncertainty. 2–3) Theory of ferromagnetism. 4) Theory of conductivity (Bloch's paper). 5) Retarded potentials in Qu. th. I think, this will not interfere with your plans. On the steamer we might speak about the details, so that our lectures fit somewhat together.[37]

One week later, Heisenberg wrote, 'Our steamer is called *Shinyo Maru*, and stops at Honolulu Aug. 20th . . . You need a Japanese visa and a 'sailing permit,' which confirms, that you have paid your income tax in U.S.A. . . . I will probably arrive in Berkeley Aug. 12th and give three lectures there. On the way . . . I will probably stop one or two days in Yellowstone Park. Are you going there too?'[38] Final arrangements between the two were concluded in letters of 19 July and 1 August; they included dinner together in Chicago before Dirac, who was starting west earlier, was to board the night train.

During July, Dirac was in Ann Arbor, lecturing at the famous Summer School in physics at the University of Michigan, one of a series that was started in 1927 by Harrison Randall, Chairman of the Physics Department. While Heisenberg was trying to reserve passage to Japan, Dirac was corresponding with Nishina in Tokyo about his forthcoming visit, and also about travel arrangements after leaving Japan.

Yoshio Nishina, their main host in Japan, graduated in Electrical Engineering from Tokyo University in 1918 (as Dirac did from Bristol five years later), and had then joined the recently founded Institute for Physical and Chemical Research (whose Japanese abbreviated name is *Riken*) in Tokyo. In 1921, *Riken* sent him abroad to improve his scientific skills. He worked for a year in Cambridge at the Cavendish Laboratory, then went to study theoretical physics in Göttingen. After that, he spent six years at Bohr's institute in Copenhagen, where he wrote his famous paper on the Compton effect with Klein. After being recalled to Japan, he returned at the end of 1928 and soon became the leading figure in Japanese nuclear and cosmic ray research.

In 1927, Nishina had written to Dirac, who was in Göttingen, to arrange to visit him:

I am leaving Copenhagen on about the 1st of July. I have not a fixed plan yet, but

I shall come to Germany and France to learn languages, then about the end of this year I shall come to England to stay for a short time. After that I shall cross to America and go back to Japan . . . I should like very much to know your plan in the near future so that I can see you somewhere and learn physics from you for some time when it is most convenient for you . . . At present we have Pauli and Wentzel here for a short visit. Prof. Bohr and Prof. Darwin have just moved to Prof. Bohr's summer villa at Tisvilde.'[39]

Nishina also wrote to Dirac twice from Hamburg, in February 1928, to say that he would spend a few months at Copenhagen (at Bohr's suggestion) before going to see Dirac at Cambridge, and that he intended to sail for America at the beginning of October.[40] Upon Dirac's replying that he would not be in Cambridge that summer, Nishina proposed to join Dirac in Leiden (where Dirac *would* be) '. . . in order to learn different things from you.' He also asked for 'a separate cover of your last paper on "the theory of electron" . . . in case you can spare one. I hope to calculate Compton-effect according to your new theory.'[41] The visit was actually made in Cambridge in fall. After sailing to America and crossing the continent, a weary Nishina wrote from Pasadena that he was sailing from San Francisco on 5 December 1928. He added, 'When I left Cambridge a month earlier, I told you that there might be a possibility of inviting you to Japan on your way back, I do not know whether it can be realized. I shall talk with some people in Japan about it.' He then asked Dirac what month would be most convenient for his visit.[42]

Dirac received two letters in July 1929 from Nishina in Tokyo in response to a request to reserve a seat on the Trans-Siberian railway. There was a potential difficulty, as Nishina wrote:

For last few days, the newspapers here have been reporting the tension in the Russo-Chinese relation with regard to the Eastern Chinese Railway, which connects South Manchuria with Central Siberia and also Vladivostock with Central Siberia, both via Harbin . . . Russians might close the frontier between China. I do not know how far these news are reliable . . . If you could not travel by way of Manchuria, you could cross to Vladivostock and then get to Central Siberia without coming into Manchuria, so long as Russians do not close Siberia to foreigners . . . In case this route were also closed, you would have to travel by steamer either via Indian Ocean or via America.[43]

In the second letter, Nishina reported that, although the Russo-Chinese troubles had not been settled, it would still be all right for Dirac to return to England by going through Vladivostock: '... there is a train connexion between Vladivostock and Moscow ... without touching Manchuria. This train runs regularly once a week and the journey takes a few days more than that through Manchuria; they say the journey is quite all right ... Looking forward very much to seeing you again at Yokohama Pier.'[44] In the end, Dirac did make his return to England via Vladivostock.

Before leaving America, the two friends visited the University of California at Berkeley. Each lectured on the morning of August 13, Dirac on the vector model (47) and Heisenberg on ferromagnetism (38). As Van Vleck remarked, the two works are related, both use the idea of the exchange force, and Heisenberg made use of Dirac's results in presenting his own lecture. In the afternoon, beginning at 4.00 p.m., Heisenberg gave a double lecture on his work with Pauli on quantum electrodynamics. The Dean, Walter M. Hart, had been sent a letter by the Physics Department Chairman asking that Dirac receive $50 and Heisenberg $150 'before the close of banking hours on Tuesday, August 13' because 'these gentlemen are sailing for Japan on the 14th'.[45]

Dirac and Heisenberg sailed as planned from San Francisco, arriving in Honolulu on 20 August 1929, where they were given a friendly welcome and a sightseeing tour by the local physicists, but there was no recognition that they were anything special, just a pair of rather youthful European scientists. They were not asked to lecture, nor to hold informal discussions.[46] Their ship left the next day and proceeded to Japan, where their visit made a considerably greater impact.

They arrived in Yokohama on 30 August, and they gave eight lectures in Tokyo during 2–7 September. On September 8, they registered at the Kanaya Hotel in Nikko, a mountain resort not far from Tokyo.[47] After that, they lectured at Kyoto Imperial University, where they were heard by both Sinitiro Tomonaga and Hideki Yukawa, who were then third-year students at the University. The latter wrote in his autobiography: 'Attending those lectures was a great stimulus to me.'[48]

The Tokyo lectures were officially sponsored by an organization called the Venus Society (*Kei mekai*), with some financial assistance from *Riken*.[49] The actual arrangements were made by Nishina and the distinguished older physicist Hantaro Nagaoka; they also made notes of the lectures for subsequent publication in the Japanese language.[50] In the preface to that publication, dated August 1931, Nishina states that lectures were held both at Tokyo Imperial University and at *Riken*. The first lecture of each speaker was supposed to be less specialized than the remainder. While the translator made some modifications of the material and added some footnotes, he tried to keep as close as possible to the original.

Tomonaga made the trip from Kyoto to Tokyo to hear the lectures, as he reported in his book *Spin Turns*. According to him, the two more general lectures were held at *Riken*, the others at the University. He continued:

Fortunately, I had read the relevant papers beforehand, so almost all the contents were known to me. But being shy and coming from Kyoto to the big city of Tokyo, I took a rear seat. One of my friends, who was a graduate of the University of Tokyo . . . , pointed out Nishina to me, and also the young Kotani and Inui, . . . who were regular members of Nishina's colloquium. He wanted me to meet them, but I hesitated. One incident I remember well was a question addressed to Heisenberg by Dirac. The question was this: Heisenberg and Pauli had used the q-number relation

$$(\nabla \cdot \mathbf{E} - 4\pi\rho)\psi = 0.$$

Dirac asked whether the zero eigenvalue of the operator is discrete or belongs to the continuum. Heisenberg had not expected that question, so he had to consider it for a while. Then he answered, 'I think it may be part of the continuum.' At the end of the University of Tokyo lectures, Professor Nagaoka gave congratulations in English to Dirac and Heisenberg for their brilliant contributions made while they were still in their twenties. 'Japanese scholars are still only studying the results of European and American achievements and our students are only taking notes. What a poor situation that is!' (48).

We conclude this account of this trip around the world with a letter that Heisenberg wrote to Dirac in December 1929 after his return to Leipzig:

Many thanks for your letter. I heard about your new paper already a few days ago from Landau (via Gamow). I think I understand the idea of your paper; it is certainly a great progress. But I cannot see yet, how the ratio of the masses etc. will come out. It seems to me already very doubtful, whether the terms of the electron (i.e. Sommerfeld formula) will not be completely changed by the interaction with the negative cells. One may hope, that all these difficulties will be solved by straight calculation of the interaction. In this case it would probably be necessary, to treat the interaction in a proper relativistic way, so

Pauli's and my paper might be of some use. I was almost sorry to send you the first paper, since the second simplifies the whole matter very much. Our present scheme is very similar to your old theory of radiation – Coulomb interaction e^2/r plus interaction between radiation and matter; the scheme is relativistically invariant in spite of e^2/r (r = space – distance).

My trip via Shanghai–Hongkong etc. was very pleasant and interesting; the best part was the trip to the biggest and best mountains, I had very good luck with the weather just the morning on Tiger Hill. In India itself it was very hot and rather rainy. Once our train went off the rails in the middle of the Jungle and people were very afraid of tigers; the tigers probably were pretty afraid too. In the Red Sea the temperature at night was 95°, the water temperature in the swimming pool 89°. On the boat I had very nice company, mostly Japanese businessmen. I did not see the Southern Cross, so you are right again. You are wrong however in the question of mating a King and a Knight with King and Castle; this is *not* possible according to the edition of 1926 of Dufresne's handbook of chess (the best book about theory of chess).[51]

4 The later years: quantum field theory and elementary particles

After the tempestuous decade of the twenties, which saw the development of the theory of quantum mechanics, its applications, and its relativistic extensions to the theory of the electron and to quantum field theory, there followed a period of slower progress in those areas, especially as seen in the work of Heisenberg and Dirac. The chief concern of their overlapping interests was quantum field theory, and this encountered both mathematical and conceptual difficulties (especially the 'infinities'), which could not be handled until the renormalization program was introduced in the 1940s (and not completely, even then). On the other hand, an ever increasing number of new particles was discovered – positron, mesotron (later muon), pion, kaon, antiproton, hyperons, excited nucleon states, etc. – whose behavior, if not whose very existence, was to be explained by quantum field theory. These new constituents of matter (mostly unexpected) required the introduction and elaboration of new types of fields in addition to those already known; their new types of interactions created mathematical difficulties that were even more severe than those which plagued quantum electrodynamics. In attempting to deal with these problems, Dirac and Heisenberg took paths that were essentially separate, although occasionally their paths crossed.

After 1933, there were fewer opportunities for the two friends to meet, as compared with the earlier 'golden age.' Before the outbreak of the war, Heisenberg went to England only twice – to Cambridge in 1934 and to Bristol in 1938; aside from those trips abroad, he went on several occasions to Copenhagen, where he met with Niels Bohr, and with his associates and visitors. Of course, between 1939 and 1945, the war interrupted virtually all communication between the scientists on opposite sides.[52] From July 1945 to January 1946, Heisenberg and other leading German scientists who were working in the Uranium Project were detained in Farm Hall near Cambridge; during this time, they were visited occasionally by some British scientists (including P. M. S. Blackett). Heisenberg went again to Britain in December 1947, this time as a free man, to lecture at the Universities of Cambridge, Edinburgh, and Bristol.[53] Only late in the 1950s, however, did Dirac and Heisenberg begin to see each other again with some regularity: e.g., at the Lindau meetings of Nobel Laureates (in 1959, 1962, and 1969); at the Solvay Conference in Brussels in 1961; at the Bohr memorial meeting in Copenhagen in 1963; the Feldafing symposium in 1965; and at conferences in Trieste in 1968 and 1972. On these occasions, they eagerly seized the opportunity to be together as in the old days, discussing recent advances in physics and their own work.

4.1 Dirac's and Heisenberg's work on quantum field theory and related topics after 1935

The starting point for Dirac's renewed attack on the problems of quantum field theory can be indicated by quoting from the 1935 edition of his book, *The Principles of Quantum Mechanics*, which ends with a few remarks on the status of quantum electrodynamics, including the following:

The foregoing theory provides a quantum electrodynamics which is a satisfactory analogue of classical electrodynamics, each of the features of classical electrodynamics having its quantum counterpart. As a description of nature, though, the theory is incomplete, as it suffers from the same limitations

Dirac (right) and Heisenberg. Summer 1962, during the Lindau Nobel Prize Winners Meeting (on the streets of Lindau).

as the classical theory. The quantum theory that we have set up also leads to an infinite mass of the electron ... It seems that some essential new physical ideas are here needed.[54]

Since the classical and the quantum theory both had difficulties in treating the point electron, Dirac began by trying to improve the classical electron theory. Specifically, he modified the interaction of the electron with the radiation field such that the infinite Coulomb self-energy was avoided (1938). A few years later, he suggested a new interpretation of quantum field theory, which made use of an indefinite metric in the mathematical space of states (1941). The problem of improving the foundations of classical electron theory continued to occupy Dirac for many years. In his attempt to modify the theory, he even proposed that one might give up the equivalence of the Heisenberg and Schrödinger pictures in quantum mechanics, which he had himself demonstrated for the non-relativistic theory about thirty years earlier! (49). At some point, he considered a possible revival of the ether concept (1951, 1953); he also played with the idea of an extended electron, whose first quantum excitation might be the muon (1962).

During this same period, Heisenberg was pursuing different, and perhaps even more ambitious goals. In a letter to Dirac in 1935, in which he thanked the latter for sending a copy of his new edition of *The Principles of Quantum Mechanics*, he said, 'I don't believe at all any more in your conjecture that the Sommerfeld fine structure constant may have something to do with the temperature concept ... Rather, I am firmly convinced that one must determine e^2/hc within the hole theory itself, in order for the theory to be sensible.'[55] However, after all his attempts to obtain the fine structure constant from hole theory failed, and he became increasingly involved in the study of cosmic ray phenomena, Heisenberg soon turned to quite different ideas.

New particles (neutrino, mesotron) and new high energy phenomena, especially the so-called explosive showers in cosmic rays, persuaded Heisenberg that it would be necessary to abandon the existing quantum field theory (1936, 1939).[56] Since the new field theories, i.e., Fermi's theory of beta decay and Yukawa's meson theory, led to even greater mathematical difficulties than quantum electrodynamics, he constructed a radically alternative description of elementary particle interactions, involving only observable quantities: S-matrix theory

144

(*Streumatrixtheorie*, 1942, 1944). However, when he was unable to find enough general conditions to determine the scattering completely, he turned to quantum field theory once again (after 1946), this time introducing new features like non-locality and non-linearity. The new theory, he hoped, would describe all elementary particles and interactions through a differential equation.

In the unified field theory that he began to develop in 1953, Heisenberg made essential use of a method due to Dirac, as he stated in the letter acknowledging the receipt of still another edition of Dirac's treatise on quantum mechanics (the fourth edition):

I have in the past years repeatedly had the experience that when one has any sort of doubt about difficult fundamental mathematical problems and their formal representation, it is best to consult your book, because these questions are treated most carefully in your book. Moreover, it will interest you that we have meanwhile made great progress with the indefinite metric in Hilbert space, which you already introduced some time ago, so that we can no longer doubt that it forms an essential part of the ultimately valid theory of the elementary particles.[57]

4.2 The last scientific exchange of Heisenberg and Dirac: the indefinite metric

In his Bakerian Lecture to the Royal Society of London, 'The Physical Interpretation of Quantum Mechanics', delivered on 19 June 1941, Dirac said about quantum field theory: 'The simplest way of developing a theory would make it apply to a hypothetical world in which the initial probability of certain states is negative, but transition probabilities calculated for this hypothetical world are found to be always positive, and it is, again, reasonable to assume that these transition probabilities are the same as those for the actual world' (50). That is, he suggested that one should keep the standard formalism of quantum field theory, but should apply a new interpretation involving negative probabilities; that would permit the removal of some of the infinities.

Although Dirac found that the good results obtained from standard quantum field theory were retained in his new version, it was criticized by Pauli, who analyzed its consequences soon after Dirac proposed it. Pauli noted that 'the interpretation of the new method is not a consistent and complete system, but consists of certain preliminary rules for

145

computing probability coefficients of radiation and collision processes,' and he went on to say that '... the theory does not give the correct dependence of the transition probabilities ... on the number of particles initially present in the different states.'[58] With his student, Josef Maria Jauch, he used Dirac's method to treat the emission of soft photons by an electron in an external field, arriving at the conclusion: 'The result is ... not satisfactory' (52).

For quantum electrodynamics, Suraj N. Gupta and Konrad Bleuler were able to develop a formalism in which states of negative probability could be introduced consistently (53). These were states of scalar photons, and they did not lead to any negative transition probabilities, owing to the supplementary condition satisfied by the electromagnetic four-vector potential. However, one could also formulate quantum electrodynamics without using states of negative norm – indeed, that was the standard method developed after the war (sometimes called the Dirac–Schwinger formalism) – so Pauli said in 1952:

As a curiosity I might mention in passing that it was not a mathematician but a physicist, namely P. A. M. Dirac, who had the idea of putting aside the axiom according to which the probabilities must lie between the numbers 0 and 1, and (retaining the other axioms) also allowing 'negative probabilities' to enter (with constant and normalized sum of all probabilities). These generalized 'probabilities' are naturally no longer interpretable as frequencies. One must also say that those far-reaching applications in physics originally foreseen by Dirac have not proven feasible. Nevertheless they are occasionally useful for auxiliary mathematical quantities which have no direct physical meaning.[59]

As a mathematical artifice, Pauli had, himself, used Dirac's negative probabilities a few years earlier, in order to regularize infinite integrals occurring in quantum field theory, but the masses of the states having negative probability were at the end of the calculation taken to be infinite, hence, unphysical (54). Pauli and Gunnar Källen found that states of negative norm also occurred in T. D. Lee's solvable field theoretical model (55), as well as in the convergent quantum field theory that Heisenberg developed, beginning in 1953. The latter theory contains a spectral function $\rho(x^2)$ (related to a mass spectrum) which satisfies the pair of conditions (56):

$$\int \rho(x^2)\,dx^2 = 0, \qquad \int \rho(x^2)x^2\,dx^2 = 0.$$

At the Pisa Conference in June 1955, Pauli claimed that these conditions implied an indefinite metric in Hilbert space. The second condition, he continued, 'seems to indicate the presence of pairs of states with a square of the rest mass ε and 0 respectively, of which one has positive, the other negative "probabilities." The result of a suitable limiting process $\varepsilon \to 0$ for such a pair of states one may call a "dipole ghost." ... an S-matrix which is unitary without exception is possible in such a theory only if transitions from normal to abnormal states are always excluded. Heisenberg claims the existence of such a general selection rule in his theory, but I am awaiting at present a further clarification of this point ...' (57).[60]

Responding to this challenge, Heisenberg studied the dipole ghost in the Lee model, and showed that, at least in the one-heavy-particle sector, the dipole ghost destroyed neither the unitarity of the S-matrix nor the probability interpretation (58). Pauli, thereupon, worked with Heisenberg on a non-linear spinor theory of elementary particles with indefinite metric, containing a dipole ghost. However, Pauli withdrew his name from a planned publication in early 1958, and Heisenberg, instead, elaborated the theory with young collaborators in Munich (59). At the beginning of 1965, after applying the non-linear spinor theory to try to determine the fine structure constant (60), Heisenberg wrote to Dirac:

I am looking forward to our meeting in Lindau next summer. By the same mail I am sending a preprint about the famous coupling constant e^2/hc which we have discussed so much in old times. As you will see from the paper, in the end the calculations are really quite simple and the numerical result is very satisfactory. The coupling constant is actually determined not within quantum electrodynamics alone but by connecting quantum electrodynamics with a general theory of elementary particles. I am convinced that this is the final solution in spite of unavoidable shortcomings of approximation methods.[61]

4.3 Opinions and outlook

The non-linear spinor theory was a unified quantum field theory intended to describe *all* elementary particles and their interactions in a single scheme. Heisenberg had a number of opportunities to discuss his theory with Dirac, beginning with the Lindau meetings of 1959 and 1962.[62] In summer 1965, i.e., at the time of publication of the calculation of the fine structure constant mentioned above, Dirac and Heisenberg were together twice: first in Lindau, then in Feldafing at a symposium on unified theories of elementary particles. Dirac always listened attentively to Heisenberg's lectures but did not say much in the public discussions. Two years after these meetings, he expressed his opinion in a letter, writing:

My main objection to your work is that I do not think your basic (non-linear field) equation has sufficient mathematical beauty to be a fundamental equation of physics. The correct equation, when it is discovered, will probably involve some new kind of mathematics and will excite great interest among the pure mathematicians, just like Einstein's theory of the gravitational field did (and still does). The existing mathematical formalism just seems to me inadequate. Maybe your theory makes the best possible use of the existing mathematics, but to see whether this is so one would have to study other possibilities, e.g., a theory based on quarks.[63]

Dirac and Heisenberg had travelled a long way together. They knew each other well, and each admired the other's work. But they were independent thinkers who often emphasized quite different ideas. To Dirac, the guiding principle of research was mathematical beauty, as he said:

It seems to be one of the fundamental features of nature that fundamental physical laws are described in terms of a mathematical theory of great beauty and power, needing quite a high standard of mathematics for one to understand it ... Just by studying mathematics we can hope to make a guess at the kind of mathematics that will come into the physics of the future ... It may well be that the next advance in physics will come about along these lines: people first discovering the equations and then needing a few years of development in order to find the physical ideas behind the equations.[64]

Dirac usually limited his attention to one problem at a time. Heisenberg, perhaps recalling his great success with quantum mechanics in 1925, preferred to make a global attack on a whole class of problems. For example, while Dirac strove mainly to improve the theory of quantum electrodynamics, Heisenberg tried to solve the whole elementary particle theory at one coup. Heisenberg thought that nature demanded a non-linear description, and despite his love for theoretical beauty, he was willing to admit mathematical tools that Dirac considered ugly. To Heisenberg, it was the physical ideas that had to be beautiful, even if their mathematical representations appeared clumsy at the outset.

Notwithstanding these differences, they generally agreed in their views on many questions of recent physics, For example, Heisenberg once said: '... it is natural that even nowadays many experimental physicists – even some theoreticians – still look for *really* elementary particles. They hope for instance that quarks, if they existed, could play this role. I think that this is an error.' To which Dirac commented: 'I am in general agreement with this point of view about elementary particles that a concept really doesn't exist, but there is a reservation. I wonder whether the electron should not be considered as an elementary particle. It may be that I am prejudiced because I have had success with the electron and no success with other particles.'[65]

The two friends were of one mind on the renormalization procedure. In one of his last public lectures, Dirac said:

... we should no longer have to make use of such illogical processes as infinite renormalization. This is quite nonsense physically, and I have always been opposed to it ... In spite of its successes, one should be prepared to abandon it completely and look on all the successes that have been obtained by using the usual forms of quantum electrodynamics with the infinities removed by artificial processes as just accidents when they give the right answers, in the same way as the successes of Bohr's theory are considerely merely as accidents when they turn out to be correct.[66]

Had Heisenberg been present, he would have nodded in agreement!

Now both of them are gone, and we have no way of knowing what they might have thought about the recent progress in the theory of elementary particles and the new problems that face it. However, we remain profoundly grateful for the ideas of Dirac and Heisenberg, which, for over half a century, enriched our physics and broke new paths towards answering its fundamental questions.

Notes

1 In a letter dated 20 November 1925, Heisenberg wrote to Dirac:

Ihre außerordentlich schöne Arbeit über Quantenmechanik hab ich mit dem größten Interesse gelesen und es kann wohl kein Zweifel sein, daß alle Ihre Resultate richtig sind, sofern man überhaupt an die neue versuchte Theorie glaubt ... Hoffentlich sind Sie nun nicht betrübt darüber, daß allerdings ein Teil Ihrer Resultate auch hier vor einiger Zeit schon gefunden werde und daher in zwei Arbeiten (die eine von Born und Jordan, die andere von Born, Jordan und mir) unabhängig in *Zeitschrift für Physik* erscheinen wird. Aber deswegen sind Ihre Resultate keineswegs unrichtiger geworden; denn einerseits gehen Ihre Ergebnisse, insbesondere was die allgemeine Definition des Differentialquotienten und den Zusammenhang der Quanten-bedingungen mit den Poissonschen Klammersymbolen anlangt, erheblich über die genannten Arbeiten hinaus, andererseits ist Ihre Arbeit auch wirklich engentlich besser und konzentrierter geschrieben als unsere Versuche hier.

(Translation, partially by Dirac, who occasionally understates Heisenberg's enthusiastic language:

I read your extraordinarily beautiful paper on quantum mechanics with the greatest interest. There can be no doubt that all your results are correct, insofar as one believes in the new, attempted theory at all ... Now I hope you are not disturbed by the fact that indeed part of your results have already been found here some time ago and are published independently here in two papers – one by Born and Jordan, the other by Born, Jordan, and me – in *Zeitschrift für Physik*. However, because of this your results by no means have become less important; on the one hand, your results, especially concerning the general definition of the differential quotient and the connection of the quantum conditions with the Poisson brackets, go considerably further than the mentioned work; on the other hand, your paper is also written really better and more concisely than our formulations given here.)

Unless otherwise stated, the letters received by Dirac are in the archives of Churchill College, Cambridge, UK. We wish to thank the Archivist of that college for providing access to them. Letters received by Heisenberg during this period are, unfortunately, not available, and their present whereabouts are unknown. Some of Heisenberg's papers were removed by the United States military authorities in 1945, and some were destroyed during an air raid on Leipzig in the fall of 1943.

2 Heisenberg to Bohr, 20 November 1925:

Heut bekam ich von Herrn Dirac ... eine Arbeit zugeschickt, in der er im wesentlichen den mathematischen Teil zur neuen Quantenmechanik auf Grund meiner Arbeit auch gemacht hat (unabhängig von Born und Jordan) ... In der Schreibweise gefällt mir manches sogar noch besser, als bei Born und Jordan.

3 Heisenberg to Dirac, 9 April 1926:

Zunächst gratuliere ich Ihnen sehr, ich war ganz begeistert als ich die Arbeit las. Ihre Teilung des Problems in die zwei Teile – Rechnen mit den 'q-numbers' einerseits – physikalische Interpretation dieser 'q-numbers' andererseits – scheint mir vollkommen dem Wesen des mathematischen Problems zu entsprechen. Bei Ihrer Behandlung des Wasserstoffproblems scheint mir auch nur noch ein kleiner Schritt zu Berechnung der Übergangs wahrscheinlichkeiten.

Dirac and Heisenberg – a partnership in science

In the last remark, Heisenberg is referring implicitly to the work of his friend, Wolfgang Pauli, in Hamburg, who had, in the meanwhile, also calculated the hydrogen spectrum, but had not been able to obtain the line intensities (13).

4 Heisenberg to Dirac, 9 April 1926:

Vor ein paar Wochen kam eine Arbeit von Schrödinger ... heraus ..., deren Inhalt meiner Ansicht nach mathematisch eng mit der Quantenmechanik zusammenhängen muß. Haben Sie sich überlegt, wie weit diese Schrödingersche Behandlung des Wasserstoffatoms mit der quantenmechanischen zusammenhämgt? Mich interessiert diese mathematische Frage deswegen besonders, weil ich glaube, daß man für die physikalische, prinzipielle Bedeutung des Formalismus sehr viel gewinnen könnte.

5 This connection had already been established by Pauli in a letter to Jordan of 12 April 1926 and independently and earlier by Schrödinger in a paper received on 18 March by *Annalen der Physik* (14).

6 Although Dirac may have seen Fermi's paper, he did not remember it when he worked on the symmetry properties of the Schrödinger function.

7 See (23), p. 506:

Die hierfür durchgeführten Rechnungen scheinen mir ein Argument dafür, daß eine kontinuierliche Interpretation des quantenmechanischen Formalismus, also auch der de Broglie–Schrödingerschen Wellen, nicht dem Wesen der bekannten formalen Zusammenhänge entsprechen würde.

8 As a particular example, Dirac showed the equivalence of the problem of diagonalizing the energy matrix in the Born–Heisenberg–Jordan theory with the problem of finding the energy eigenvalues of Schrödinger's wave equation. To accomplish this proof of the equivalence, he invented the singular delta function.

9 From Dirac's letter to Jordan, 24 December 1926. This time, Dirac had beaten the Göttingen group by two weeks, since Jordan's paper (26) was received on 18 December 1926.

10 See Heisenberg (27), p. 172:

Diese Ungenauigkeit ist der eigentliche Grund für das Auftreten statistischer Zusammenhänge in der Quantenmechanik. Ihre mathematische Formulierung gelingt mittels der Dirac–Jordanschen Theorie. Von den so gewonnenen Grundsätzen ausgehend wird gezeigt, wie die makroskopischen Vorgänge aus der Quantenmechanik heraus verstanden werden können.

11 At the Fifth Solvay Conference, held from 24 to 29 October 1927, besides the senior quantum physicists (Planck, Einstein, Born, and Bohr), also the younger generation (de Broglie, Dirac, Heisenberg, Pauli, and Schrödinger) participated. Of these, only de Broglie and Schrödinger presented reports, but many of Heisenberg's results were incorporated in the reports of Born

151

(and Heisenberg) and of Bohr. Bohr had already presented his report a little earlier at the Como Congress in September (28). It contained the first statement of Bohr's principle of complementarity. However, it was at the Solvay Conference that this principle – which, together with Heisenberg's uncertainty relation provided the basis of the 'Copenhagen interpretation' of quantum mechanics – was discussed for the first time.

12 Some historical studies of Dirac's relativistic electron theory are: Donald Franklin Moyer, *Amer. J. Phys.* **49**, 944, 1055, 1120 (1981); Helge Kragh, *Arch. Hist. Exact Sci.* **24**, 31 (1981). Dirac has discussed his electron equation in *Recollections of an Exciting Era*, ref. (9), and also in *Directions in Physics*, Wiley, New York (1978) and in 'The Prediction of Antimatter,' H. R. Crane Lecture, Univ. of Michigan, Ann Arbor (17 April 1978).

13 See also Dirac's review article, The origin of quantum field theory, in *The Birth of Particle Physics*, pp. 39–55, eds. Laurie M. Brown and Lillian Hoddeson, Cambridge Univ. Press (1983). For a historical discussion see Joan Bromberg in *History of Twentieth Century Physics*, ref. (9), p. 243.

14 Heisenberg to N. Bohr, 31 March 1928:

Dirac's Arbeiten habe ich auch sehr bewundert; aber ich finde es sehr beunruhigend, daß eine scheinbar so geschlossene Theorie wie die Dirac'sche eine so schlimme Lücke aufweist, wie die Übergänge von positiver zu negativer Energie.

On 26 November 1929, Dirac wrote to Bohr from Cambridge, suggesting that one could avoid the difficulty of negative energy electron states by assuming that most of those states were already occupied by electrons, so that the Pauli principle would forbid transitions to them. He added:

It seems reasonable to assume that not all the states of negative energy are occupied, but that there are a few vacancies or 'holes.' . . . These holes I believe to be the protons.

(See Moyer, reference quoted in note 12 above, part 2, for the complete text of Dirac's letter and Bohr's reply.) Dirac published his hole theory in 1930 (37).

15 Heisenberg to Jordan, 25 June 1928:

Dirac hat hier nur seine bisherige Theorie vorgetragen, samt einer hübschen Begründung dafür, daß die Differentialgleichungen linear in $\partial/\partial x_\mu$ sein muß. Die berühmten Schwierigkeiten hat er nicht lösen können; ich habe eingehend darüber mit D. gesprochen; ich persönlich möchte glauben, daß man versuchen muß, die Unsymmetrie in $+e$ und $-e$ (m und M) irgendwie hereinzubekommen. Ein starkes Argument dafür ist ja auch, daß vermutlich das Verhältnis m/M mit der Feinstrukturkonstante zusammenhängt. Dirac ist aber anderer Ansicht und hofft noch auf widerspruchsfreie Theorie für ein einzelnes Elektron.

16 Heisenberg to Bohr, 23 July 1928:

Sehr viel unglücklicher bin ich über die Frage nach der relativistischen Formulierung und über die Inkonsequenz der Dirac-Theorie. Dirac war hier und hielt uns einen sehr schönen Vortrag über seine ja wirklich geniale Theorie. Er sieht aber ebensowenig wie wir anderen einen Weg aus der Schwierigkeit $\rightarrow e\ -e$ und die Schwierigkeit ist bei genauerer Überlegung doch viel ernster, als es mir anfangs schien. Z.B. sollten die spontanen Strahlungsübergänge $+mc^2 \rightarrow -mc^2$ wohl viel häufiger vorkommen, als irgend welche anderen spontanen Übergänge. Andererseits sieht man bei Anwendung der Dirac-Theorie auf andere Fragen, z.B. Streuung des Lichtes, bes. bei freien Elektronen, daß die ominösen Übergänge doch ein integrierender Bestandteil der Dirac-Theorie sind. Also ich find' die gegenwärtige Lage ganz absurb und hab' mich deshalb, quasi aus Verzweiflung, auf ein anderes Gebiet, das des Ferromagnetismus, begeben.

He had submitted his theory of ferromagnetism already two months earlier for publication (38).

17 Before proposing his famous linear relativistic electron equation, Dirac tried (in April 1926) to make a 'relativity quantum mechanics' by putting time and space on the same quantum mechanical footing, i.e., letting them both be represented by q-numbers. He applied the resulting theory to Compton scattering, obtaining a result that was satisfactory for photon wavelengths that were not too short, but that failed to agree with subsequent experiments at higher energy, partly because his theory did not include the electron spin (12).

18 Heisenberg to Pauli, 4 November 1926:

Ich selbst hab' ein wenig über die Theorie des Ferromagnetismus, der Leitfähigkeit und ähnliche S ... ereien nachgedacht. Die Idee ist die: Um die Langevinsche Theorie des Ferromagn[etismus] zu brauchen, muß man eine große Kopplungskraft zwischen den spinnenden Elektronen annehmen (es drehen ja *nur diese* sich). Diese Kraft soll, wie beim Helium, von der Resonanz indirekt geliefert werden. Ich glaube, man kann allgemain beweisen: Parallelstellung der Spinvektoren gibt stets *kleinste* Energie ... Ich hab' das Gefühl ..., daß die im Prinzip zu einer Deutung des Ferromagnetismus ausreichen könnte.

19 See the report of R. J. Eden and J. C. Polkinghorne, Dirac in Cambridge, in *Aspects of Quantum Theory*, pp. 1–5, eds. A. Salam and E. P. Wigner, Cambridge Univ. Press (1972).

20 Dirac, from 1928 on, worked on developing a relativistic quantum field theory (which he had initiated earlier (35)), in addition to his work on the relativistic electron theory. See papers (39) and (40).

21 See the letter of Heisenberg to Pauli of 23 February 1927 in *Wolfgang Pauli: Wissenschaftlicher Briefwechsel mit Bohr, Einstein, Heisenberg u.a./Scientific Correspondence, Volume I: 1919–1929*, pp. 376–81, eds. A. Hermann, K. v. Meyenn and V. F. Weisskopf, Springer Verlag, New York-Heidelberg-Berlin (1979).

22 Pauli and Jordan had already considered the case of 'symmetric statistics,' i.e., boson matter fields (42). Pauli thought that the antisymmetric case (fermions) would not be more difficult.

23 Pauli to Dirac, 17 February 1928:

> In dieser Verbindung habe ich auch darüber nachgedacht, in *welchem Verhältnis Ihre neue Quantentheorie des Elektrons zu Ihrer früheren Quantenelektrodynamik ... steht.*

('In this connection, I have also thought about *the relation of your new quantum theory of the electron to your earlier quantum electrodynamics.*')

24 See Victor Weisskopf, Growing up with field theory, in *The Birth of Particle Physics*, reference quoted in note 13, pp. 56–81.

25 Heisenberg to Bohr, 1 March 1929 (from *S.S. George Washington* of the United States Line):

> Besonders gespannt bin ich, was Du über das Verhältnis zwischen Quantentheorie und Relativitätstheorie sagen wirst. Gelegentlich der Arbeit mit Pauli hab ich auch viel über diese prinzipiellen Dinge nachgedacht, doch ohne viel Erfolg; die Arbeit von Pauli und mir ist zunächst im besten Falle ein formaler Fortschritt, vielleicht nicht einmal das, weil die Dirac-Schwierigkeiten natürlich ungelöst bleiben ... Ob ich in Amerika zu irgendwelchen Arbeiten komm', scheint mir noch unklar; zu meiner Freude hörte ich, daß Dirac auch in den nächsten Monaten drüben sein wird, da kann man doch ein wenig über Theorie diskutieren.

26 Heisenberg to Bohr, 20 December 1929:

> Pauli und meine Elektrodynamik hat sich weiter ganz außerordentlich vereinfacht, die früheren Zusatzglieder sind alle ganz überflüssig geworden und bis auf formale Unschönheiten scheint mir dieses Wechselwirkungsproblem jetzt in Ordnung; aber es bleibt halt eine sehr graue Theorie, solange die Diracsche Schwierigkeit ungelöst ist. Dirac hat ja über die $+e$ Angelegenheit eine neue Arbeit geschrieben; Dur wirst sie jedenfalls auch kennen. Ich bin aber recht skeptisch, weil die Protonenmasse gleich der Elektronenmass herauskommt, soviel ich sehe.

27 On 26 January 1934, Bohr wrote from Copenhagen to Yoshio Nishina, who had returned to Tokyo in 1929, '... the striking confirmation which this (Klein–Nishina) formula has obtained became soon the main support for the essential correctness of Dirac's theory when it was apparently confronted with so many grave difficulties.'

28 See Laurie M. Brown & Donald F. Moyer, Lady or tiger? – The Meitner–Hupfeld effect and Heisenberg's neutron theory, *Amer. J. Phys.* **52**, 130–6 (1984).

29 J. H. Van Vleck, Travels with Dirac in the Rockies, in *Aspects of Quantum Theory* (note 19), pp. 7–16.

30 Heisenberg to Dirac, 13 February 1928, from Leipzig.

31 Heisenberg to Dirac, 23 January 1929, from Leipzig.

32 Heisenberg to Dirac, from Cambridge, Mass., undated.

33 Van Vleck, note 28.

34 Roundy, interview published in *Wisconsin State Journal*, 31 April 1929. The complete text is given by S. S. Schweber in *Relativity, Groups and Topology II*, pp. 54–5, eds. B. S. De Witt & R. Stora, North-Holland, Amsterdam (1984).

35 Heisenberg to Dirac, 24 May 1929, from Chicago.

36 Heisenberg to Dirac, 17 June 1929, from Chicago.

37 Heisenberg to Dirac, 5 July 1929, from Chicago; quoted in full.

38 Heisenberg to Dirac, 11 July 1929.

39 Nishina to Dirac, 11 June 1927, from Copenhagen. Dirac was also well-acquainted with Y. Sugiura, another Japanese physicist doing research in Europe at that time. Sugiura wrote to Dirac from Göttingen on 26 September 1927 to say that he wanted to visit Cambridge in October to listen to Dirac's lectures. In 1929, when Dirac and Heisenberg visited Japan, they met both Nishina and Sugiura, who had returned in the meantime.

40 Nishina to Dirac, 10 February 1928, from Hamburg.

41 Nishina to Dirac, 25 February 1928, from Hamburg.

42 Nishina to Dirac, 3 December 1928, from Pasadena.

43 Nishina to Dirac, 16 July 1929, from Tokyo.

44 Nishina to Dirac, 30 July 1929, from Tokyo.

45 The information on the Berkeley visit is in the R. T. Birge letters in the Bancroft Library of the University of California at Berkeley. We are very grateful to Professor Manuel G. Doncel for acquainting us with this material and a few other suggestions.

46 A brief anecdotal account of the Dirac–Heisenberg visit to Honolulu is given by Van Vleck, note 28. A more detailed and documented version is by S. F. Tuan, 'Dirac and Heisenberg in Hawaii.' We thank Professor Tuan for a copy of his unpublished manuscript.

47 A photocopy of the hotel register is given by Tuan, note 46.

48 Hideki Yukawa, *Tabibito* (The Traveler), p. 71, translated by L. M. Brown and R. Yoshida, World Scientific, Singapore (1982).

49 The Venus Society was established in 1918 on the basis of a one million yen grant by Tetsuma Akaboshi, with the purpose of encouraging new invention and discovery, supporting research and publication, translating foreign books, etc. Nagaoka was a member of its board of trustees.

50 The lectures were published by the Venus Society under the title, *Various Problems in Quantum Theory*. Nishina was the translator, assisted by M. Kotani and T. Inui. The publication date was 20 April 1932. The titles of the lectures were given in both English and Japanese as follows:

Laurie M. Brown and Helmut Rechenberg

Heisenberg:
I. The Indeterminary-Relations and the Physical Principles of the Quantum Theory.
II. Theory of Ferromagnetism.
III. Theory of Conduction.
IV. Retarded Potential in the Quantum Theory.

Dirac:
I. The Principle of Superposition and the Two-Dimensional Oscillator.
II. The Basis of Statistical Quantum Mechanics.
III. Quantum Mechanics of Many-Electron Systems.
IV. Relativity Theory of Electron.

51 Heisenberg to Dirac, 7 December 1929.

52 The only word of Dirac that Heisenberg had during the war was very indirect. On 25 March 1943, Josef Teszler, a Roumanian Jew who had married Dirac's sister, Beatrice, in 1937, wrote to Heisenberg from Amsterdam to ask for advice in what he described as 'our threatening situation' (*unserer bedrohlichen Lage*). In the same letter, he wrote, 'We occasionally hear about Paul via the Red Cross and are happy that he is well.' (*Wir hören von Paul ab und zu durchs Rote Kreuz und sind froh, daß es ihm gut geht.*) Heisenberg, having sought out various possibilities, replied to Teszler, in a letter dated 20 July 1943, that he could not yet offer any specific help, but hoped to contact him a little later when visiting Holland. Although Heisenberg did go to Holland in the fall of 1943, he seems not to have made contact with Teszler. The latter wrote again on 2 February 1944, this time from Budapest, urgently requesting a statement confirming the Aryan descent of his wife, which Heisenberg promptly sent.

53 On that occasion, Heisenberg met many former colleagues and his old teacher, Max Born. However, he did not meet Dirac, who was in Princeton. On 12 January 1948, Dirac's article, 'Quantum Theory of Localizable Dynamical Systems', was received by the *Physical Review*, sent from the Institute for Advanced Study. The first two references in this work are to papers of Heisenberg.

54 Dirac, ref. (46), second edition, 1935, pp. 296–7.

55 Heisenberg to Dirac, 27 March 1935:

Ich glaube gar nicht mehr an Ihre Vermutung, daß die Sommerfeldsche Feinstrukturkonstante etwas mit dem Temperaturbegriff zu tun hätte ... Vielmehr bin ich fest davon überzeugt, daß man schon in der Löchertheorie e^2/hc bestimmen muß, bevor die Theorie vernünftig werden kann.

Dirac and Heisenberg – a partnership in science

56 For a recent historical account, see David C. Cassidy, Cosmic ray showers, high energy physics, and quantum field theories: Programmatic interactions in the 1930s, in *Historical Studies in the Physical Sciences* **12**, 1 (1981).

57 Heisenberg to Dirac, 27 February 1958:

> Ich habe in den vergangenen Jahren immer wieder die Erfahrung gemacht, daß man bei den schwierigen, grundsätzlichen mathematischen Problemen und ihrer formalen Darstellung am besten in Ihrem Buch nachschaut, wenn man irgendwelche Zweifel hat, weil diese Fragen in Ihrem Buch am sorgfältigsten behandelt sind. Übrigens wird es Sie interessieren, daß wir mit der indefiniten Metrik in Hilbertraum, die ja seinerzeit von Ihnen eingeführt wurde, so daß wir nicht mehr daran zweifeln können, daß sie ein wesentlicher Bestandteil der endgültigen Theorie der Elementarteilchen ist.

58 See ref. (51). This is an extended version of a lecture that Pauli gave at Purdue University in June 1942. In it, Pauli took Dirac's proposal seriously and analyzed it in detail. In a footnote to this article, Pauli calls it 'a continuation of the earlier report, *Rev. Mod. Phys.* **13**, 203 (1941).' The latter article, entitled 'Relativistic Field Theories of Elementary Particles,' was an elaboration of part of Pauli's report prepared for the 1939 Solvay Conference – as the first part of a joint report with Heisenberg on the properties of elementary particles. (The Solvay Conference was cancelled due to the outbreak of war. Recently, Heisenberg's report was included in *Volume B of his Collected Works*, p. 346, eds. W. Blum, H.-P. Dürr & H. Rechenberg, Springer-Verlag, Berlin (1984).) Pauli thus considered his analysis of Dirac's new method of field quantization as part of a new series of papers on quantum field theory.

59 W. Pauli, Wahrscheinlichkeit un Physik, *Dialectica* **8**, 283 (1954):

> Als Kuriosum möchte ich nebenbei anführen, daß nicht ein Mathematiker, sondern ein Physiker, nämlich P. A. M. Dirac, auf die Idee kam, daß Axiom, wonach die Wahrscheinlichkeiten zwischen 0 und 1 gelegene Zahlen sein müssen, fallen zu lassen und (unter Beibehaltung der übrigen Axiome) auch 'negative Wahrscheinlichkeiten' (mit konstanter und normierter Summe aller Wahrscheinlichkeiten) zuzulassen. Diese verallgemeinerten 'Wahrscheinlichkeiten' sind natürlich nicht mehr als Häufigkeiten interpretierbar. Dementsprechend hat sich auch deren von Dirac ursprünglich vorgesehene weitgehende Anwendung in der Physik als nicht durchführbar erwiesen. Dennoch sind sie aber zuweilen nützlich für mathematische Hilfsgrößen, die keine direkte physikalische Bedeutung haben.

60 Pauli and Villars (ref. 54) had noticed earlier that their regularization procedure implied these same mass spectral conditions.

61 Heisenberg to Dirac, 19 January 1965.

62 One of the authors (H.R.) was present at both Lindau meetings and observed that Dirac and Heisenberg remained together at every opportunity.

63 Dirac to Heisenberg, 6 March 1967.
64 Dirac, ref. 61, p. 53.
65 Heisenberg, Development of concepts in the history of quantum theory, discussion in *The Physicist's Conception of Nature*, pp. 264–75 (on p. 273), ed. Jagdish Mehra, Reidel, Dordrecht-Boston (1973).
66 Dirac, The origin of quantum field theory, note 13, on p. 100.

References

(1) Interview with C. F. von Weizsäcker, 9 June 1963, *Archive for History of Quantum Physics* (AHQP).
(2) P. A. M. Dirac, When a golden age started (introduction to Heisenberg's lecture), in *From a Life of Physics – Evening Lectures at the International Centre for Theoretical Physics, Trieste, Italy*, p. 32, IAEA, Vienna (1968).
(3) P. A. M. Dirac, *Conversation with Mehra*, quoted in J. Mehra & H. Rechenberg, *The Historical Development of Quantum Theory*, Vol. 4, p. 5, Springer-Verlag, New York (1982).
(4) W. Heisenberg, Über die quantentheoretische Umdeutung kinematischer und mechanischer Beziehungen, *Z. Phys.* **33**, 879–893 (receiver 29 July 1925, published in issue No. 12 of 18 September 1925).
(5) See ref. (3), p. 5.
(6) P. A. M. Dirac, The fundamental equations of quantum mechanics, *Proc. Roy. Soc. London* **A109**, 642–53 (received 7 November 1925, published in issue No. **A752** of 1 December 1925).
(7) M. Born & P. Jordan, Zur Quantenmechanik, *Z. Phys.* **34**, 858—888 (received 27 September 1925, published in issue No. 11/12 of 28 November 1925).
(8) M. Born, W. Heisenberg & P. Jordan, Zur Quantenmechanik. II, *Z. Phys.* **35**, 557–615 (received 16 November 1925, published in issue No. 8/9 of 4 February 1926).
(9) P. A. M. Dirac, Recollections of an exciting era, in *History of Twentieth Century Physics*, pp. 109–46, ed. C. Weiner, Academic Press, New York (1977).
(10) P. A. M. Dirac, Quantum mechanics and a preliminary investigation of the hydrogen atom, *Proc. Roy. Soc. London* **A110**, 561–79 (received 22 January 1926, published in issue No. **A755** of 1 March 1926).
(11) P. A. M. Dirac, The elimination of the nodes in quantum mechanics, *Proc. Roy. Soc. London* **A111**, 281–305 (received 27 March 1926, published in issue No. **A757** of 1 May 1926).

(12) P. A. M. Dirac, Relativity quantum mechanics with an application to Compton scattering, *Proc. Roy. Soc. London* **A111**, 405–23 (received 29 April 1926, published in issue No. **A758** of 2 June 1926).

(13) W. Pauli, Über das Wasserstoffspektrum vom Standpunkt der neuen Quantenmechanik, *Z. Phys.* **36**, 336–63 (received 17 January 1926, published in issue No. 5 of 27 March 1926).

(14) E. Schrödinger, Über das Verhältnis der Heisenberg–Born–Jordanschen Quantenmechanik zu der meinen, *Ann. d. Phys.* (4) **79**, 734–56 (received 18 March 1926, published in issue No. 8 of 4 May 1926).

(15) E. Schrödinger, Quantisierung als Eigenwertproblem (Erste Mitteilung), *Ann. d. Phys.* (4) **79**, 361–76 (received 27 January 1926, published in issue No. 4 of 13 March 1926).

(16) P. A. M. Dirac, On the theory of quantum mechanics, *Proc. Roy. Soc. London* **A112**, 661–77 (received 26 August 1926, published in issue No. **A762** of 1 October 1926).

(17) E. Fermi, Sulla quantizzazione del gas perfetto monoatomico, *Rend. R. Acad. Lincei* **3**, 145–9 (presented at the meeting of 7 February 1926); Zur Quantelung des idealen einatomigen Gases, *Z. Phys.* **36**, 902–12 (received 24 March 1926, published in issue No. 11/12 of 11 May 1926).

(18) G. Wentzel, Die mehrfach periodischen Systeme in der Quantenmechanik, *Z. Phys.* **37**, 80–94 (received 27 March 1926, published in issue No. 1/2 of 22 May 1926).

(19) W. Heisenberg & P. Jordan, Anwendung der Quantenmechanik auf das Problem der anomalen Zeemaneffekte, *Z. Phys.* **37**, 263–77 (received 16 March 1926, published in issue No. 4/5 of J June 1926).

(20) W. Heisenberg, Mehrkörperproblem und Resonanz in der Quantenmechanik, *Z. Phys.* **38**, 411–26 (received 11 June 1926, published in issue No. 6/7 of 10 August 1926).

(21) W. Heisenberg, Über die Spektra von Atomsystemen mit zwei Elektronen, *Z. Phys.* **39**, 499–518 (received 24 July 1926, published in issue No. 7/8 of 26 October 1926).

(22) W. Heisenberg, Quantenmechanik, *Naturwissenschaften* **14**, 989–94 (presented on 26 September 1926, published in the issue of 5 November 1926).

(23) W. Heisenberg, Schwankungserscheinungen und Quantenmechanik, *Z. Phys.* **40**, 501–6 (received 6 November 1926, published in issue No. 7 of 20 December 1926).

(24) W. Heisenberg, Mehrkörperprobleme und Resonanz in der Quantenmechanik. II, *Z. Phys.* **41**, 239–67 (received 22 December 1926, published in issue No. 4/5 of 14 February 1927).

(25) P. A. M. Dirac, The physical interpretation of the quantum dynamics,

Proc. Roy. Soc. London **A113**, 621–41 (received 2 December 1926, published in issue No. **A765** of 1 January 1927).

(26) P. Jordan, Über eine neue Begründung der Quantenmechanik, *Z. Phys.* **40**, 809–38 (received 18 December 1926, published in issue No. 11/12 of 18 January 1927).

(27) W. Heisenberg, Über den anschaulichen Inhalt der quantentheoretischen Kinematik und Mechanik, *Z. Phys.* **43**, 172–98 (received 26 March 1927, published in issue No. 3/4 of 31 May 1927).

(28) N. Bohr, The quantum postulate and the recent development of atomic theory, in *Atti del Congresso Internazionale dei Fisici, Como, Settembre 1927*, Bologna 1928, pp. 565–88.

(29) P. A. M. Dirac, The quantum theory of the electron, *Proc. Roy. Soc. London* **A117**, 610–24 (received 2 January 1928, published in issue No. **A778** of 1 February 1928); The quantum theory of the electron. Part II, *Proc. Roy. Soc. London* **A118**, 351–60 (received 2 February 1928, published in issue No. **A779** of 1 March 1928); Über die Quantentheorie des Elektrons, *Phys. Zeitschr.* **29**, 561–3 (summary of a lecture presented at the *Leipziger Universitätswoche, 18–23 June 1923*, published in issue No. 16 of 15 August 1928).

(30) C. G. Darwin, The wave equations of the electrons, *Proc. Roy. Soc. London* **A118**, 654–80 (received 6 March 1928, published in issue No. **A780** of 2 April 1928).

(31) W. Gordon, Die Energieniveaus des Wasserstoffatoms nach der Diracschen Quantentheorie des Elektrons, *Z. Phys.* **48**, 11–14 (received 23 February 1928, published in issue No. 1/2 of 5 April 1928).

(32) P. A. M. Dirac, Quantum singularities in the electromagnetic field, *Proc. Roy. Soc. London* **A133**, 60–72 (received 29 May 1931, published in issue No. **A821** of 1 September 1931).

(33) C. D. Anderson, Energies of cosmic-ray particles, *Phys. Rev.* **41**, 405–21 (received 28 June 1932, published in issue No. 4 of August 1932).

(34) P. M. S. Blackett and G. P. S. Occhialini, Some photographs of the tracks of penetrating radiation, *Proc. Roy. Soc. London* **A139**, 699–718 (received 7 February 1933, published in issue No. **A839** of 8 March 1933).

(35) P. A. M. Dirac, On the quantum theory of emission and absorption of radiation, *Proc. Roy. Soc. London* **A114**, 243–65 (received 2 February 1927, published in issue No. **A767** of 1 March 1927); On the quantum theory of dispersion, *Proc. Roy. Soc. London* **A114**, 710–28 (received 4 April 1927, published in issue No. **A769** of 2 May 1927).

(36) W. Heisenberg, *Physics and Beyond*, p. 93, Harper and Row, New York (1971) (translation by A. J. Pomerans of *Der Teil und das Ganze*, R. Piper Verlag, Munich (1969)).

(37) P. A. M. Dirac, A theory of electrons and protons, *Proc. Roy. Soc. London* **A126**, 360–5 (received 6 December 1929, published in issue No. **A801** of 1 January 1930).

(38) W. Heisenberg, Zur Theorie des Ferromagnetismus, *Z. Phys.* **49**, 619–36 (received 20 May 1928, published in issue No. 9/10 of 16 July 1928).

(39) P. A. M. Dirac, Relativistic quantum mechanics, *Proc. Roy. Soc. London* **A136**, 453–64 (received 4 March 1932, published in No. **A829** of 2 May 1932).

(40) P. A. M. Dirac, V. A. Fock and B. Podolsky, On quantum electrodynamics, *Phys. Z. der Sowjetunion* **3**, 64–72 (1932).

(41) P. Jordan & E. Wigner, Über das Paulische Äquivalenzverbot, *Z. Phys.* **47**, 631–51 (received 26 January 1928, published in issue No. 9/10 of 16 March 1928).

(42) P. Jordan & W. Pauli, Zur Quantenelektrodynamik ladungsfreier Felder, *Z. Phys.* **47**, 151–73 (received 7 December 1927, published in issue No. 3/4 of 4 February 1928).

(43) W. Heisenberg & W. Pauli, Zur Quantendynamik der Wellenfelder, *Z. Phys.* **56**, 1–61 (received 19 March 1929, published in issue No. 1/2 of 8 July 1929); Zur Quantentheorie der Wellenfelder. II, *Z. Phys.* **59**, 168–90 (received 7 September 1929, published in issue No. 3/4 of 2 January 1930).

(44) O. Klein & Y. Nishina, Über die Streuung von Strahlung durch freie Elektronen nach der neuen relativistischen Quantendynamik von Dirac, *Z. Phys.* **52**, 853–68 (received 30 October 1928, published in issue No. 11/12 of 9 January 1929); Y. Nishina, Die Polarisation der Comptonstreuung nach der Diracschen Theorie des Elektrons, *Z. Phys.* **52**, 869–77 (received 23 November 1928, published in issue No. 11/12 of 9 January 1929).

(45) W. Heisenberg, *The Physical Principles of the Quantum Theory*, Univ. of Chicago Press, Chicago (1930).

(46) P. A. M. Dirac, *The Principles of Quantum Mechanics*, Clarendon Press, Oxford (1930).

(47) P. A. M. Dirac, Quantum mechanics of many-electron systems, *Proc. Roy. Soc. London* **A123**, 714–33 (received 12 March 1929, published in issue No. **A792** of 6 April 1929).

(48) S. Tomonaga, Spin Turns (Chuokon ronsha), 1974.

(49) P. A. M. Dirac, Quantum theory of localizable dynamical systems, *Phys. Rev.* (2) **73**, 1092–1103 (received 12 January 1948, published in issue No. 9 of 1 May 1948).

(50) P. A. M. Dirac, The physical interpretation of quantum mechanics, *Proc. Roy. Soc. London* **A180**, 1–40 (received 23 September 1941, published in issue No. **A980** of 18 March 1942).

(51) W. Pauli, On Dirac's new method of field quantization, *Rev. Mod. Phys.* **15**,

175–207 (published in issue No. 3 of July 1943).

(52) W. Pauli & J. M. Jauch, On the application of Dirac's method of field quantization to the problem of emission of low frequency photons, *Phys. Rev.* (2) **65**, 255–6 (abstract of a paper presented at the New York Meeting of the American Physical Society, 14–15 January 1944, published in issue No. 7/8 of 1/15 April 1944).

(53) S. N. Gupta, Theory of longitudinal photons in quantum electrodynamics, *Proc. Roy. Soc. London* **A63**, 681–91 (received 20 June 1949, published in issue No. **A367** of 1 July 1950); K. Bleuler, Eine neue Methode zur Behandlung der longitudinalen und skalaren Photonen, *Helv. Phys. Acta* **23**, 567–86 (received 10 June 1950, published in issue No. 5 of 1 September 1950).

(54) W. Pauli & F. Villars, On the invariant regularization in relativistic quantum field theory, *Rev. Mod. Phys.* **21**, 434–44 (received 10 May 1949, published in issue No. 3 of July 1949).

(55) G. Källen & W. Pauli, On the mathematical structure of T. D. Lee's model of a renormalizable field theory, *Kgl. Danske Videnskab. Selskab. Mat.-Phys. Medd.* **30**, No. 7, 3–23 (received 15 April 1955).

(56) W. Heisenberg, Zur Quantisierung nichtlinearer Gleichungen, *Nachr. Akad. Wiss. Göttingen, Math.-Phys.-Chem. Abt. 1953*, 111–27 (presented on 6 November 1953, published in issue No. 8).

(57) W. Pauli, Remarks on problems connected with the renormalization of quantum fields, *Nuovo Cimento (10)* **4**, Suppl., 703–10 (paper presented at the Pisa Conference, 12–18 June 1955, published in issue No. 2 of 1956).

(58) W. Heisenberg, Lee model and quantization of non linear field equations, *Nuclear Physics* **4**, 532–63 (received 14 July 1957).

(59) H. P. Dürr, W. Heisenberg, H. Mitter, S. Schlieder & K. Yamazaki, Zur Theorie der Elementarteilchen, *Z. für Naturforschung* **14a**, 441–85 (received 3 March 1959, published in issue No. 5/6).

(60) H. P. Dürr, W. Heisenberg, H. Yamamoto & K. Yamazaki, Quantum electrodynamics in the nonlinear spinor theory and the value of Sommerfeld's fine-structure constant, *Nuovo Cimento (10)* **38**, 1120–42 (received 29 January 1965, published in issue No. 3 of 1 August 1965).

(61) P. A. M. Dirac, The evolution of the physicist's picture of nature, *Scientific American* **208**, No. 5, 45–53 (published in May 1963).

13

Dirac's magnetic monopole and the fine structure constant

William J. Marciano and
Maurice Goldhaber
Brookhaven National Laboratory*

In 1931, P. A. M. Dirac[1] published his seminal work, 'Quantized Singularities in the Electromagnetic Field', which brought together for the first time quantum mechanics and magnetic monopoles. Seventeen years later, he further clarified and reinforced his earlier results in a second classic study,[2] 'The theory of Magnetic Poles'. The motivation behind that body of work was Dirac's desire to explain electric charge quantization (i.e., why all observed charges are integral multiples of the electron's charge e) and to understand the value of the fine structure constant

$$\alpha \equiv e^2/4\pi\hbar c \simeq 1/137. \tag{1}$$

Indeed, finding a simple mathematical explanation for the measured value of α was considered one of the outstanding theoretical problems of that era.

Dirac showed that, by introducing magnetic charge in addition to electric charge, Maxwell's equations became more symmetric with respect to electric–magnetic interchange. In fact, their vacuum dual symmetry $\mathbf{E} \to \mathbf{B}$, $\mathbf{B} \to -\mathbf{E}$ would also be present in matter if there existed magnetic currents k^μ which similarly transformed into electric currents j^μ under duality ($j^\mu \to k^\mu$, $k^\mu \to j^\mu$). However, describing

* Supported by the Department of Energy, U.S.A.

163

magnetic monopoles required potentials with string singularities extending from the poles out to infinity, the so-called Dirac Strings[2]. Of course, singularities are generally mathematical artifacts in physics; so Dirac demanded that string singularities be unobservable in any quantum process. That requirement led to the now famous Dirac Quantization Condition*

$$\frac{QM}{4\pi} = \frac{n}{2}\hbar c, \qquad n = \text{integer}, \qquad (2)$$

where Q and M denote electric and magnetic charges. The appearance of Planck's constant on the right-hand side of this relationship shows its quantum mechanical origin. In fact, (2) can be viewed as a half-integral quantization condition on the angular momentum stored in the electromagnetic fields of electric and magnetic charges.[3] Of course, any formalism that relies on singular potentials is subject to some skepticism, and such was the case for the Dirac Quantization Condition. However, over the years, it has been carefully scrutinized and rederived in a number of very different approaches; so its validity is no longer seriously questioned.[4]

The implications of (2) are truly remarkable. Taking the electron's charge e to be the smallest (unconfined) unit of electric charge implies

$$M_{\min} = \frac{2\pi}{e}, \qquad \text{Dirac Magnetic Charge} \qquad (3)$$

for the value of the smallest (non-zero) unit of magnetic charge.

Inserting that value back into (2) gives an electric charge quantization condition

$$Q = ne, \qquad n = \text{integer}. \qquad (4)$$

So, the existence of magnetic charge implies electric charge quantization; thus, fulfilling the first half of Dirac's goal. However, employing $e^2/4\pi \equiv \alpha \simeq 1/137$, (3) indicates

$$M_{\min} \simeq \frac{137}{2}e. \qquad (5)$$

* In keeping with more modern convention, we have scaled Dirac's original electric and magnetic units by $1/\sqrt{4\pi}$. Also, following (2), we will subsequently employ natural units $\hbar = c = 1$.

The minimum non-zero magnetic charge is predicted to be much larger than the basic unit of electric charge. It seems that Nature chose to put a big asymmetry in electric–magnetic duality. Whereas quantum electromagnetic processes involving electrically charged particles are well described by perturbative quantum electrodynamics (QED) with a small expansion parameter $\alpha \simeq 1/137$, magnetic monopole interactions are governed by

$$M_{min}^2/4\pi \simeq 137/4 \simeq 34,$$

a very strong coupling. Therefore, the physics of magnetic monopoles should be quite different from what we are accustomed to.

Unfortunately, little can be said about the properties of Dirac's original magnetic monopole, other than it should have a M_{min}/r^2 magnetic field with $M_{min} \simeq 137e/2$. Of course, such a large field suggests a big mass; but the classical self-energy of a point magnetic monopole is infinite (as is a point electric charge), so its mass is not predicted. Over the years, numerous experiments have searched for magnetic monopoles using mainly the anticipated large magnetic coupling as guidance. Although, from time to time, candidate monopole events were reported, none have stood up under closer scrutiny or better statistics.[5]

Modern day grand unified theories (GUTS)[6] also predict the existence of magnetic monopoles.[7-8] In those theories, electric charge quantization follows directly from the fact that they possess a simple compact gauge group structure (e.g., SU(5), SO(10), E_6, etc.); so there can only be one gauge coupling g_{GUT}. Electric charges are related to that parameter by simple group weighting factors. Magnetic monopoles exist as topologically stable soliton-like field configurations with short-distance core structure. Their existence depends on the presence of a U(1) subgroup in the unbroken symmetry of the theory. For example, the spontaneous breaking of $G = $ SU(5), SO(10) or E_6 down to the standard model, $SU(3)_C \times SU(2)_L \times U(1)$ (i.e., QCD $\times$ Electroweak theory) at mass scale m_{GUT} gives rise to a family of magnetic monopoles.[8,9-11] The lightest monopole carries one unit of ordinary Dirac magnetic charge $2\pi/e$, as well as (screened) color magnetic charge. That explains how the Dirac Quantization Condition applies to fractionally charged quarks. It is the total angular momentum stored in ordinary electromagnetic fields plus color fields of the quark–magnetic monopole system that must be quantized in half-integer units of $\hbar$. The mass of the lightest (stable) magnetic monopole in GUT models is finite (due to the core structure) and generally predicted to be (in the Prasad–Sommerfeld limit)[8,10-12]

$$m_{\text{monopole}} = m_{\text{GUT}}/\alpha_{\text{GUT}}, \tag{6}$$

where $\alpha_{\text{GUT}} \equiv g_{\text{GUT}}^2/4\pi$ is the unified gauge coupling value at the unification mass scale m_{GUT}. Since proton decay can be mediated by gauge bosons $X^{\pm 4/3}, Y^{\pm 1/3}$ in GUT models and those particles have mass m_{GUT}, we can use present experimental bounds on the proton lifetime[13]

$$\tau_{p \to e^+ \pi^0} \geqslant 2.5 \times 10^{32} \, \text{yr} \tag{7a}$$

to infer[14]

$$m_{\text{GUT}} \geqslant 10^{15} \, \text{GeV}. \tag{7b}$$

So, GUT monopoles are expected to be very heavy and, therefore, have production thresholds way beyond the capabilities of any presently conceivable high energy accelerator. It seems that the only way magnetic monopoles may be observable is as leftover remnants of the very early universe.

Besides carrying ordinary and color magnetic charge and being very massive, GUT magnetic monopoles have another important property. They are able to catalyze proton decay, Monopole $+ p \to$ Monopole $+ e^+ + X$ with fairly large cross-sections.[15-16] That feature provides further useful guidance for magnetic monopole searches[5] and means that ongoing proton decay experiments[13] which employ massive underground detectors are also sensitive to very small monopole fluxes. So, it seems that Dirac's magnetic monopole finds a natural setting in GUTS and, within that framework, it exhibits more fantastic properties[9] than one possibly could have imagined in 1931.

Although Dirac was successful in explaining electric charge quantization by introducing magnetic monopoles, he was, nevertheless, disappointed[17] in not being able to shed light on the value of α. However, our attitude towards α as a fundamental number has changed over the years. We now view it (as well as all other couplings) as an effective charge which is screened (or antiscreened) by quantum vacuum polarization effects. Its value depends on the distance (or energy scale μ) probed, as well as the renormalization prescription employed. The usual fine structure constant α is defined through Thomson scattering in the limit of zero momentum transfer and, thus, corresponds to a value as seen at infinite distance. At shorter distances (higher energy probes), the effective electromagnetic coupling $\alpha(\mu) \equiv e^2(\mu)/4\pi$ grows. It satisfies a simple (perturbative) renormalization group equation

$$\mu \frac{d}{d\mu} \alpha(\mu) \equiv \beta(\alpha) = b_0 \alpha^2(\mu) + \text{higher order terms}, \qquad (8a)$$

where[18]

$$b_0 = \frac{2}{3\pi} \sum_f Q_f^2 \theta(\mu - m_f) - \frac{7}{2\pi} \theta(\mu - m_W). \qquad (8b)$$

The summation in (8b) is over all electrically charged fermions (quarks and leptons) which screen the charge, while the second term describes $W^\pm$ boson antiscreening effects. (Thresholds are incorporated with a step-function as is appropriate for the $\overline{\text{MS}}$ (modified minimum subtraction) prescription.[18]) If other, as yet undiscovered, elementary charged particles exist, their contribution must be included at energies μ above their masses.

Integrating (8) and employing known particle masses, as well as data on $e^+ e^- \rightarrow$ hadrons to incorporate strong interaction effects, one finds[18] that the $\overline{\text{MS}}$ electromagnetic coupling at $\mu = m_W \simeq 83$ GeV (i.e., distances $\simeq 2 \times 10^{-16}$ cm) is*

$$\alpha^{-1}(m_W) \simeq 127.7 \pm 0.3. \qquad (9)$$

The 7% increase, due to fermion screening effects, in going from infinite distance ($\mu = 0$) to 2×10^{-16} cm is supported by high energy $e^+ e^-$ scattering experiments. So, we see that the number $\alpha \simeq 1/137$ is not particularly special or fundamental; at least no more so than the $1/127.7$ in (9). As a matter of fact, within the framework of GUTS, interactions become more symmetric at short distances; so, one would expect a viable 'derivation' of $\alpha(\mu)$ from first principles to occur at some very large mass scale such as m_{GUT} or $m_{\text{Planck}} = (G_N)^{-1/2} \simeq 10^{19}$ GeV. For example, in many popular GUTS (SU(5), SO(10), E_6, etc.), the $SU(3)_C \times SU(2)_L \times U(1)$ couplings $\alpha_i(\mu)$, $i = 1, 2, 3$ of the standard model become unified[19] (very nearly equal) at $m_{\text{GUT}} > 10^{15}$ GeV. For

---

* At $\mu = 0$, the $\overline{\text{MS}}$ coupling is related to the usual fine structure constant via[18] $\alpha^{-1}(0) = \alpha^{-1} + 1/6\pi$.

example, in the $\overline{\text{MS}}$ scheme, one finds

$$\alpha_1^{-1}(m_{\text{GUT}}) = \alpha_2^{-1}(m_{\text{GUT}}) - \frac{1}{6\pi} = \alpha_3^{-1}(m_{\text{GUT}}) - \frac{1}{4\pi} \qquad (10a)$$

$$\alpha^{-1}(m_{\text{GUT}}) = \tfrac{5}{3}\alpha_1^{-1}(m_{\text{GUT}}) + \alpha_2^{-1}(m_{\text{GUT}}), \qquad (10b)$$

where the small differences in (10a) arise in the usual $\overline{\text{MS}}$ definitions.[18] (They could be absorbed into a redefinition of the couplings.) A potential program for 'deriving' α would be to find a symmetry or stability argument that determines $\alpha_i(m_{\text{GUT}})$ and, consequently, $\alpha(m_{\text{GUT}})$ via (10b). Then, assuming all charged particles with masses $< m_{\text{GUT}}$ were known, one could integrate (8) down to $\mu = 0$ and obtain α.

Given the above discussion, one should naturally ask: At what value of μ does Dirac's quantization condition

$$\frac{e(\mu)M_{\min}(\mu)}{4\pi} = \frac{1}{2} \qquad (11)$$

hold? Although that issue is not completely settled, we expect it to hold for all μ. (In modern terminology, it is renormalization group invariant.) In fact, for a given definition of the renormalized charge $e(\mu)$, (11) should be used to define the corresponding renormalized magnetic charge and to determine its dependence on μ. That reciprocal relationship would seem to be a necessary condition for any consistent theory, since the dielectric constant and magnetic permeability of the vacuum determine electric and magnetic screening and Lorentz invariance of the vacuum requires[20] that the

dielectric constant = (magnetic permeability)$^{-1}$.

Accepting the fact that $e(\mu)$ grows while $M_{\min}(\mu)$ decreases with increasing μ (at least up to m_{GUT}), then, for some appropriate definition of the renormalized charges, there may be a mass scale where they are nearly equal and duality becomes manifest. In GUT models, m_{GUT} could be the right scale for such a symmetry restoration. We would like to speculate that, for an appropriate definition of the renormalized charge* $e'(\mu)$, exact electric–magnetic duality (in all fields) is achieved at $\mu = m_{\text{GUT}}$, the primed electric charge of the $X^{\pm 4/3}$ gauge boson is equal

---

* The primed couplings differ from the usual $\overline{\text{MS}}$ definition.

to the primed magnetic charge of the lightest magnetic monopole

$$\tfrac{4}{3}e'(m_{\text{GUT}}) = \frac{2\pi}{e'(m_{\text{GUT}})}, \quad \text{Duality Conjecture.} \tag{12a}$$

The duality conjecture in (12a), taken together with Dirac's quantization condition, determines the unification couplings

$$\alpha'(m_{\text{GUT}}) = \tfrac{3}{8} \tag{12b}$$

$$\alpha_i'(m_{\text{GUT}}) = 1, \quad i = 1, 2, 3. \tag{12c}$$

Of course, since we haven't defined the primed couplings, there is little real content in Equations (12). Indeed, we will take Equations (12) as a starting definition of the primed couplings, hoping that they will eventually emerge from some as yet hidden symmetry,[21] as fixed points, or perhaps from dynamical considerations. To motivate that possibility, we note from (6) that, in the Prasad–Sommerfeld limit,[12] the lightest magnetic monopole is degenerate with $m_{\text{GUT}} = m_X$ to lowest order in $\alpha_i'(m_X)$. We conjecture that the degeneracy may, in fact, be nearly exact

$$m_X = m_{\text{GUT}} = m_{\text{monopole}} \tag{13}$$

and helping in defining $\alpha'(m_{\text{GUT}})$. Such a situation might occur in superstring models,[22] where all large mass scales are nearly degenerate[23]

$$m_X = m_{\text{GUT}} = m_{\text{compactification}} \simeq m_{\text{string}} \approx m_{\text{Planck}}. \tag{14}$$

For the primed couplings to have perturbative meaning, they must be relatively small at low $\mu \simeq m_W$ and presumably not very different than the experimental $\overline{\text{MS}}$ coupling values

$$\alpha_3(m_W) = 0.10 \pm 0.01 \tag{15a}$$

$$\alpha_2(m_W) = 0.036 \pm 0.002 \tag{15b}$$

$$\alpha_1(m_W) = 0.0167 \pm 0.0003 \tag{15c}$$

$$\alpha(m_W) = 1/(127.7 \pm 0.3). \tag{15d}$$

In fact, they should be related by

$$\frac{1}{\alpha_i'(\mu)} \simeq \frac{1}{\alpha_i(\mu)} - C, \quad i = 1, 2, 3 \tag{16a}$$

$$\frac{1}{\alpha'(\mu)} \simeq \frac{1}{\alpha(\mu)} - \tfrac{8}{3}C, \tag{16b}$$

where C accounts for the different renormalization prescriptions (definitions). That being the case, the $\alpha_i(\mu)$ and $\alpha_i'(\mu)$ will have the same perturbative β-*functions* up to $O(\alpha_i^3)$. To examine the approximate conditions necessary to have $\alpha'(m_{\mathrm{GUT}})$ evolve from $\simeq 3/8$ at unification to $\simeq \alpha(m_W) \simeq 1/128$, we can integrate (8), making the simplifying assumption that all as yet undiscovered fermions have mass m_F. Then, using (9) and (12) as input, we find that $\alpha'(m_W) \simeq \alpha(m_W)$ requires

$$125 \simeq \frac{11}{6\pi} \ln\left(\frac{m_{\mathrm{GUT}}}{m_W}\right) + \frac{2}{3\pi} \sum_F Q_F^2 \ln\left(\frac{m_{\mathrm{GUT}}}{m_F}\right), \tag{17}$$

where the sum is over all fermions beyond the usual three generations. Anticipating that m_{GUT} may be near the Planck mass in such a scenario, we take $m_{\mathrm{GUT}} \simeq 10^{18}$ GeV and find that (17) is satisfied if

$$\sum_F Q_F^2 \ln\left(\frac{1 \times 10^{18}\ \mathrm{GeV}}{m_F}\right) \simeq 487. \tag{18}$$

Since each complete generation of quarks and leptons gives $\sum_F Q_F^2 = 8/3$, one way to fulfill this requirement is to have five additional relatively light generations with $m_F \simeq 137$ GeV. So, the discovery of five more fermion generations could be construed as providing indirect evidence for the duality conjecture and would be suggestive of strong coupling at unification.

A large value of m_{GUT} near 10^{18} GeV, as we have envisioned, most naturally occurs in supersymmetric GUT models.[24] Furthermore, supersymmetry may be needed for exact duality to occur. In those scenarios, there are fermion partners of gauge bosons and scalar partners of ordinary fermions that contribute to charge screening; so, not too many generations are needed to have $\alpha'(\mu)$ run from 3/8 at m_{GUT} to $\simeq 1/128$ at m_W. As a realistic example, we focus on a supersymmetric E_6 model[25] with three relatively light $\underline{27}$-plets of fermions and their supersymmetric scalar partners. Each $\underline{27}$-plet contains an ordinary fermion generation as well as an additional color triplet quark and three new leptons. Such a model is a candidate effective theory for $\mu < m_{\mathrm{GUT}}$ in $E_8 \times E_8$ superstring theory.[26] Assuming that all supersymmetric partners of gauge bosons and fermions as well as the additional members of the 27-plets have mass of $O(m_W)$ and $\alpha_i'(m_{GUT}) = \frac{8}{3}\alpha'(m_{GUT}) = 1$ at $m_{GUT} \simeq 10^{18}$ GeV, we integrate the 3 generation supersymmetric 2 loop β-functions[27-28] for the $\alpha_i'(\mu)$ and find

$$\alpha_3'(m_W) = 0.144 \tag{19a}$$

$$\alpha_2'(m_W) = 0.040 \tag{19b}$$

$$\alpha_1'(m_W) = 0.0173 \tag{19c}$$

$$\alpha'(m_W) = 1/121. \tag{19d}$$

These values are systematically larger than the experimental $\overline{MS}$ couplings in (15) and suggest the relationship

$$\frac{1}{\alpha_i'(\mu)} \simeq \frac{1}{\alpha_i(\mu)} - 2.5, \qquad i = 1, 2, 3. \tag{20}$$

So, at low energies, the primed and $\overline{MS}$ couplings are not so different; but, at unification, $\alpha_i'(m_{GUT}) = 1$ while $\alpha_i(m_{GUT}) \simeq 0.286$. It might appear that the $\overline{MS}$ couplings are better for perturbative expansions, since they are smaller. However, higher order corrections expressed in terms of those couplings are probably large. For example, the first term in the $\alpha_3(\mu)$ β-function vanishes for 3 supersymmetric 27-plets and one is left with[28]

$$\mu\frac{d}{d\mu}\alpha_3(\mu) = \frac{6}{\pi^2}\alpha_3^3 + \frac{9}{8\pi^2}\alpha_3^2\alpha_2 + \frac{3}{8\pi^2}\alpha_3^2\alpha_1 + b_{333}\alpha_3^4 + \cdots, \tag{21}$$

where b_{333} depends on the renormalization scheme. Although b_{333} has not been fully calculated in supersymmetric theories, based on the pure QCD calculation,[29] we would guess that in the $\overline{MS}$ scheme $b_{333} \gtrsim 39/\pi^2$, which is too large to ignore. One could eliminate the b_{333} term by defining a new expansion parameter

$$\frac{1}{\alpha_3''(\mu)} \equiv \frac{1}{\alpha_3(\mu)} - \frac{b_{333}}{(6/\pi^2)}, \tag{22}$$

which could then be viewed as a kind of optimal perturbative coupling. Using our educated guess that $b_{333} \simeq 39/\pi^2$ for this model, (22) becomes

$$\frac{1}{\alpha_3''(\mu)} \simeq \frac{1}{\alpha_3(\mu)} - 2, \tag{23}$$

which is very similar to the relationship in (20). It suggests that the $\alpha_i'(\mu)$ defined by our duality condition may well be optimal expansion parameters.

In conclusion, we have shown that, in GUT models with many relatively light fermions and scalars, such as the 3 generation supersymmetric E_6 example, the gauge couplings may well be unified near the Planck scale $\simeq 10^{19}$ GeV and be of $O(1)$ there. That being the case, effective electric and magnetic couplings would both be of similar magnitude, thereby, suggesting a dual symmetry. The occurrence of strong coupling at unification may signal a phase transition, such as the opening up of tiny extra dimensions as in Kaluza–Klein or superstring scenarios.[22,26] Unfortunately, it would also render perturbative analysis near unification dubious at best.

Dirac's magnetic monopole has inspired the creative imaginations of several generations of physicists. It has found a place in GUTS, Kaluza–Klein theories[30] and, most recently, in superstring models.[31] In each case, it has provided insights into the theory and led to interesting new predictions for magnetic monopole phenomenology. Unfortunately, magnetic monopoles have yet to be detected experimentally. Perhaps Nature made them hard to find, so we would learn to unravel and appreciate her fundamental laws during our long pursuit for the elusive Dirac magnetic monopole.

References

1 P. A. M. Dirac, *Proc. Roy. Soc.* **A133**, 60 (1931).

2 P. A. M. Dirac, *Phys. Rev.* **74**, 817 (1948).

3 M. N. Saha, *Indian J. Phys.* **10**, 145 (1936); *Phys. Rev.* **75**, 1968 (1949).

4 P. Goddard & D. Olive, *Rep. Prog. Phys.* **41**, 1357 (1978).

5 G. Giacomelli, *La Rivista del Nuovo Cimento* **7**, 1 (1984).

6 H. Georgi & S. L. Glashow, *Phys. Rev. Lett.* **32**, 438 (1974); P. Langacker, *Phys. Rep.* **72C**, 185 (1981).

7 G. 't Hooft, *Nucl. Phys.* **B79**, 276 (1974); A. M. Polyakov, *JETP Lett.* **20**, 194 (1974).

8 C. P. Dokos & T. T. Tomaras, *Phys. Rev.* **D21**, 2940 (1980).

9 J. Preskill, *Ann. Rev. Nucl. and Part. Sci.* **34**, 461 (1984).

10 C. Gardner & J. Harvey, *Phys. Rev. Lett.* **52**, 879 (1984).

11 A. Schellekens & C. Zachos, *Phys. Rev. Lett.* **50**, 1242 (1983).

12 M. Prasad & C. Sommerfeld, *Phys. Rev. Lett.* **35**, 760 (1975); E. Bogomolnyi, *Sov. J. Nucl. Phys.* **24**, 449 (1976).

13 IMB Collaboration, H. S. Park, G. Blewitt *et al.*, *Phys. Rev. Lett.* **54**, 22 (1985).

14 M. Goldhaber & W. J. Marciano, *Comm. on Nucl. and Part. Physics* (1986).

15 V. A. Rubakov, *Nucl. Phys.* **B203**, 311 (1982); C. G. Callan, *Phys. Rev.* **D26**, 2058 (1982).

16 W. Bernreuther & N. S. Craigie, *Phys. Rev. Lett.* **55**, 2555 (1985).

17 P. A. M. Dirac, *Int. J. of Theor. Phys.* **17**, 235 (1978).

18 W. Marciano, *Phys. Rev.* **D20**, 274 (1979); W. Marciano & A. Sirlin, *Phys. Rev. Lett.* **46**, 163 (1981); W. Marciano & A. Sirlin, *Proc. of the Second Workshop on Grand Unification, Ann Arbor* (1981); W. Marciano, *Phys. Rev.* **D29**, 580 (1984).

19 H. Georgi, H. Quinn & S. Weinberg, *Phys. Rev. Lett.* **33**, 451 (1974).

20 N. K. Nielsen, *Am. J. Phys.* **39**, 1171 (1981).

21 P. Goddard, J. Nuyts & D. Olive, *Nucl. Phys.* **B125**, 1 (1977); C. Montonen & D. Olive, *Phys. Lett.* **72B**, 117 (1977); D. Olive in *Proc. of the Monopoles in Quantum Field Theory*, eds. N. Craigie, P. Goddard & W. Nahm, World Scientific, Singapore (1981).

22 J. H. Schwarz, *Phys. Rep.* **89**, 223 (1982); *Superstrings*, ed. J. H. Schwarz, World Scientific, Singapore (1985); E. Witten, *Nucl. Phys.* **B258**, 75 (1985).

23 M. Dine & N. Seiberg, *Phys. Rev. Lett.* **55**, 366 (1985); V. Kaplunovsky, *Phys. Rev. Lett.* **55**, 1036 (1985).

24 S. Dimopoulos, S. Raby & F. Wilczek, *Phys. Rev.* **D24**, 1681 (1981).

25 F. Gursey, P. Ramond & P. Sikivie, *Phys. Lett.* **60B**, 177 (1976); R. Barbieri & D. Nanopoulos, *Phys. Lett.* **91B**, 369 (1980).

26 D. J. Gross, J. Harvey, E. Martinec & R. Rohm, *Phys. Rev. Lett.* **55**, 502 (1985); P. Candelas, G. Horowitz, A. Strominger & E. Witten, *Nucl. Phys.* **B258**, 46 (1985); M. Dine, V. Kaplunovsky, M. Mangano, C. Nappi & N. Seiberg, *Nucl. Phys.* **B259**, 549 (1985).

27 D. R. T. Jones, *Phys. Rev.* **D25**, 581 (1982).

28 W. J. Marciano, to be published (1986).

29 O. Tarasov, A. Vladimirov & Z. Zharkov, *Phys. Lett.* **93B**, 429 (1980).

30 D. Gross & M. Perry, *Nucl. Phys.* **B226**, 29 (1983); R. Sorkin, *Phys. Rev. Lett.* **51**, 87 (1983).

31 X.-G. Wen & E. Witten, *Nucl. Phys.* **261**, 651 (1985).

14

Magnetic monopoles and the halos of galaxies

F. Hoyle
Cockley Moor, Dockray, England

1 Introduction

At the height of the great economic depression of the early nineteen-
thirties, I was fortunate to be awarded a scholarship to the University of
Cambridge, with all expenses munificently paid by the education
authorities in my native County of Yorkshire. I came up first to
Cambridge in early October 1933 with the intention of studying physics
and chemistry, but I soon discovered that the professors whose courses I
particularly wished to attend gave their lectures under the auspices of
the Faculty of Mathematics. This circumstance led me to change my
studies to the topics and lectures associated with the Mathematical
Tripos, a switch for which I was then ill-prepared but which I have never
since regretted. After two rather torrid years of it, I came, at last, in
October 1935 to the lectures on quantum mechanics given by Paul
Dirac.

To this point, all the courses I had attended had been at specified
times. Now, at a more advanced level, the times were mostly by
arrangement between class and lecturer. You were simply informed by
the University listing that you were to call on various professors in their
college rooms at specified hours in order to fix the times of lectures by
mutual agreement. So it came about that, one morning in October 1935,
I found myself walking into St. John's, Dirac's College. His rooms, in

those days, were in the far right-hand corner of Second Court (as you headed towards the River Cam). When I reached the appropriate staircase, it was obvious from the noise which set of rooms they were. Knocking at the outer door and then poking my nose inside, I found a veritable mob already assembled there. Only about a half were concerned, as I was, with the Mathematical Tripos. The others were research students from physics and chemistry, and there was also a considerable sprinkling of older people from abroad. It was my first sight of Dirac, then thirty-four years old. Young as this might seem, he was already a Cambridge professor of four years' standing and a Nobel Laureate of two years' standing, and it was four years since he had published his first paper on magnetic monopoles (*Proc. Roy. Soc.* **A133**, 60 (1931)). In a subsequent paper (*Phys. Rev.* **74**, 817 (1948)), the earlier discussion is broadened into an action formulation, with the equations given generally instead of being based on special examples.

It seems that Dirac began his investigation of magnetic monopoles with the hope of deducing the fine structure constant,

$$\frac{e^2}{hc} = \frac{1}{137} \quad \text{approximately,} \qquad (1.1)$$

with h standing for the Planck constant divided by 2π. Towards the end of the first paper, he expresses disappointment that this objective is not achieved. However, there is a remarkable consolation prize. If even one magnetic pole of strength g exists, then, for consistency in quantum mechanics, it is essential that the electron charge be related to g by

$$g = n\frac{hc}{2e} = \frac{137}{2} ne, \qquad (1.2)$$

where n is an integer. Hence, the electron charge is quantised in the sense that, while one could imagine a number of different electrons, differences of charge can arise only from changes of n, there can be no continuum of electron charge. Reciprocally, since electrons exist, magnetic pole strengths must also be quantised. In modern discussions of magnetic poles, it is usual to set $n = 1$ in (1.2), and this will be done from here on.

Dirac suggested an inversion of the argument, namely that, because from empirical experience the electron has a unique charge, magnetic poles probably exist in order to force the empirical result to be true. This inversion has lain fallow for almost half a century, until with the emergence of modern grand unification theories physicists have become convinced fairly generally that Dirac's perception was correct. In this article, I shall show that there is also a considerable body of astronomical data which point to the same conclusion.

It is well worthwhile for every young physicist, to whom the nineteen-thirties may seem almost as remote as the age of Pericles, to take a look at Dirac's 1931 paper, especially to read its very first page where Dirac explains that progress in theoretical physics depends on the recruitment of more advanced mathematics. Not more advanced in the sense of adding epicycles to mathematical systems already in use, but more advanced in the sense of bringing-in hitherto abstract ideas in mathematics. How incredibly true this has proved to be! The first page from the 1931 paper illustrates Dirac's propensity for stating the most profound thoughts in an exceedingly short space. Speaking from personal experience, the trouble was that you read through those vital passages so quickly that it was quite hard, at the time, to appreciate their wholly fundamental significance. Richard Feynman remarks that the entire program for his development of path integrals is contained in a single page of Dirac's book *Quantum Mechanics*, page 125 of my much-thumbed second edition; and, on page 72 of the same edition, there is a passage which might almost have served as the abstract for Laurent Schwarz' treatise on distribution theory (Volumes 1 and 2, Hermann et Cie, Paris (1950–1)).

By 1938–9, I had risen as it were from being a late addition to the mob of October 1935 to being a research student of Paul Dirac, himself. There are anecdotes I could tell from that period, but, regretfully, I must desist, since in this article – like Thursday's child – I still have far to go. Yet, I must mention a remark of Dirac which had a profound effect on my career. One day, apropos of physics in general, not, I hope, of my personal prospects, he said (I am not sure I have the words exactly right, but I will put them in quotes, nevertheless):

'In 1926, people who were not particularly good could do important work, but nowadays even people who are very good cannot find important problems to solve'.

As always with Dirac, this remark proved to be far-sighted, since, over the next decade, theoretical physics was to pass through a depressingly static phase that did not pick up appreciably until the discovery of the

Lamb shift with its attendant development of renormalisation theory. In 1939, the exciting decades of particle physics through the sixties and seventies were still a long way into the future. Although I enjoyed working in theoretical physics, Dirac's implied advice was not to be ignored, and, by 1940, I had turned my attention to astronomy, a subject in which many problems were then becoming ripe for solution. I am happy that this switch now permits me, more than forty years later, to bring what appear to be significant arguments to bear concerning the existence of magnetic poles, so returning full circle to an earlier episode in my life and to what may well prove to have been one of Dirac's most profound glimpses of reality.

2 A model for the missing mass in galaxies

There is compelling evidence for the existence of massive, approximately spherical halos extending outward from the centres of spiral galaxies to distances of 50 kpc or more.[1-5] Observations of gas motions by CO emission and by 21 cm HI emission yield circular velocities $V(r)$ that are substantially independent of the central distance, r, to $r \approx 50$ kpc. In our own galaxy, Allen[6] gives $V \approx 200$ km s^{-1}, independent of r, beyond 1 kpc from the centre. From the relation

$$V^2 = \frac{GM(r)}{r},\qquad (2.1)$$

where $M(r)$ is the total mass interior to distance r, one infers $M(r)$ proportional to r for $r > 1$ kpc in the Galaxy, implying that the density ρ varies proportionally to r^{-2}, with

$$\rho(r) \approx \frac{\bar{\rho}}{3}\left(\frac{R}{r}\right)^2,\qquad (2.2)$$

R being a parameter of the order of the radius of the halo and $\bar{\rho}$ being a mean density given by

$$\frac{4\pi}{3}R^3\bar{\rho} \approx M(R),\qquad (2.3)$$

with $M(R)$ the total mass of the galaxy. Using (2.1) at $r = R$, and

177

eliminating $\bar{\rho}$ and $M(R)$ from the above equations, one obtains

$$\rho(r) \approx \frac{V^2}{4\pi G r^2}. \qquad (2.4)$$

The nature of the halo material, which comprises most of the mass of the galaxy, is unknown, since the halo material is unseen. It has been speculated that the material might be primordial neutrinos with a rest mass of 10 to 50 eV, but recent experimental work[7] puts this suggestion in considerable doubt. Faint stars of small mass have also been considered as a possible halo material, but this too is an uneasy proposal, since halos of stars even of small masses would probably have been detected, which they have not, by modern highly sensitive photon-counting devices. The further possibility suggests itself that magnetic monopoles in the halos constitute this so-called missing mass, somewhat along the lines suggested by Cabrera,[8] in assigning the local missing mass of the solar neighbourhood to monopoles. The association of missing mass with monopole halos has also been considered by Kolb *et al.*[9] and by Salpeter *et al.*[10] in preprints which are mainly concerned with imposing constraints on the idea. We shall consider the question of constraints at a later stage, after the present model has been developed a little further.

The proportionality $\rho(r) \propto r^{-2}$ in (2.2) was written down above in order that V given by (2.1) be independent of r. What one has to wonder is the meaning of such a proportionality. Recently,[11] I investigated the possibility that the halo material is in radial orbits directed in-and-out from the galactic centre, but avoiding a strict singularity at $r = 0$ by letting the line orbits miss the exact centre by a small distance r_{min}, say, r_{min} being a variable parameter from one galaxy to another. The distribution of halo material was taken to be spherically symmetric with all elements of the material moving on their outward journeys to the same distance R from the centre. Thus, $r = R$ was the outer boundary of the halo, with R taken to be 100 kiloparsecs or more in typical cases.

With the halo structure in a steady state, one can then solve for $V(r)$ as a series of approximations. In the first order, one obtains $V = $ constant, as was considered above. In second order, however, one obtains (for r not too close to R),

$$V(r) = \frac{\text{constant}}{(\ln R/r)^{\frac{1}{6}}}. \qquad (2.5)$$

Over a limited range of r, the dependence on $\ln R/r$ can be represented as a power law, viz

$$V(r) = \text{constant}(r)^{\frac{1}{6}\ln R/r_0},\qquad(2.6)$$

applicable for r reasonably close to r_0. From (2.1), we then have $M(r)$ proportional to

$$r^{1+\frac{1}{3}\ln R/r_0},\qquad(2.7)$$

and from

$$4\pi r^2 \rho(r) = \frac{\mathrm{d}M}{\mathrm{d}r},\qquad(2.8)$$

the proportionality of $\rho(r)$ on r is easily seen to be modified from $\rho \propto r^{-2}$ to

$$\rho \propto r^{-2}\cdot r^{\frac{1}{3}\ln R/r_0}.\qquad(2.9)$$

It is now relevant that the data given by Allen[6] was, itself, somewhat approximate, and that more recent work shows that $V(r)$ does, indeed, increase slowly with r, as required by (2.6). Thus, Burstein *et al.*[12] give $V(r) \propto r^{\sim 0.15}$ at large distances from the centres of S_c galaxies, from which they conclude that $\rho(r) \propto r^{-1.7\pm0.1}$ at such distances, which agrees closely with (2.9) at distances of $r \approx r_0 \approx R/3$.

I am impressed by this close agreement, partly for the personal reason that I obtained (2.5) ahead of knowing the results of Burstein *et al.*, and partly for the objective reason that the simple radial model for the motions of halo material really does look as though it must be correct, clearly implying that the material has been ejected by explosion from the galactic centre. If, indeed, the material consists of monopoles we have the further implication that at galactic centres there are 'machines' which produce monopoles in enormous quantities, 10^{13} times the mass of the Sun in typical cases, a result of profound importance both for astronomy and for theoretical physics.

3 The monopole flux

By applying the model to our own galaxy, the flux of monopolies at the Earth can be estimated with sufficient accuracy from (2.4). Writing m_m for the monopole mass and making up the density $\rho(r)$ at distance r from the galactic centre largely from monopoles yields a monopole number density $n_m(r)$ given by

$$n_m(r) = \frac{\rho(r)}{m_m} \approx \frac{V^2}{4\pi G r^2 m_m}. \tag{3.1}$$

For the total number $N_m(r)$ of monopoles interior to r, one, thus, has

$$N_m(r) = M(r)/m_m \approx \frac{V^2 r}{G m_m}. \tag{3.2}$$

Since the monopoles move in the gravitational field of the galaxy, their velocity must be of order V, and this will also be the order of their speed relative to a galactic star system such as the solar system. Hence, the flux of monopoles relative to a star system at distance r from the galactic centre is given per steradian per unit area per unit time simply by multiplying $n_m(r)$ by $V/4\pi$, viz

$$\Phi_m(r) = \text{Flux of monopoles} \approx \frac{V^3}{(4\pi)^2 G r^2 m_m}. \tag{3.3}$$

Putting $V = 200$ km s^{-1}, $r = 10$ kpc, gives a flux

$$5 \times 10^{-14}(m_{Pl}/m_m)\,\text{cm}^{-2}\,\text{s}^{-1}\,\text{sr}^{-1}, \tag{3.4}$$

when the monopole mass is expressed in terms of the Planck mass, m_{Pl}, defined by

$$m_{Pl} = \left(\frac{hc}{2G}\right)^{\frac{1}{2}} = \left(\frac{ge}{G}\right)^{\frac{1}{2}} = 1.54 \times 10^{-5}\,\text{gram} \approx 10^{19}\,\frac{\text{GeV}}{c^2}. \tag{3.5}$$

Here, the Dirac monopole charge is written as $g = hc/2e = 137e/2$, and it is interesting to note that the Planck mass can be simply expressed in terms of the pole strengths of the three inverse square forces, Gm_{Pl}^2 being the geometric mean of g^2 and e^2.

Equation (2.1) yields $M(r) \approx 4.7 \times 10^{11}\,M_\odot$ at $r = 50$ kpc, whereas Allen[6] gives the smaller value of $1.4 \times 10^{11}\,M_\odot$ for the mass of *visible* matter in the galaxy. The mass density $\rho_m(r)$ of monopoles at $r \cong 8$ kpc, the usually adopted distance of the solar system from the galactic centre, is given by (2.4) and is $\sim 10^{-24}$ gram cm^{-3}, similar to the mass density of the interstellar gas, but less by a factor of about 4 than the mass density of the stars of the solar neighbourhood.

At first sight, there seems to be a discrepancy here, for, how can the monopoles have a smaller local mass density than the visible stellar material and yet have a greater mass in total? The explanation is that the visible material is concentrated towards the galactic plane into a disk, whereas the monopole distribution has been taken in the above discussion to be spherical. The possibility that the monopole distribution may be in the form of an oblate spheroid with some concentration towards the galactic plane cannot be excluded. If this were so, both the local mass density and the flux of the monopoles would be increased according to the degree of the concentration, otherwise, nothing in principle would be changed.

4 The magnetic fields of galaxies

The identification of halo material with monopoles has, so far, depended on the elimination of other possibilities, rather than on the positive properties of the monopoles, themselves. This negative and somewhat unsatisfactory aspect of the argument will be removed in the present section. Spiral galaxies have large-scale magnetic fields with intensities of order 10^{-6} gauss towards the central plane, defined by their visible material and directed generally in the sense of the rotation of the visible material. The source of such fields has for long been a mystery, without hope of solution according to conventional ideas. The basic difficulty of the conventional position is well-illustrated by the following simple problem.

A wire of negligible resistance in the form of a circle of radius a contains a battery of electromotive force $E^{(e)}$. How long does the battery need to operate in order that the magnetic field throughout a region of dimension a surrounding the wire shall rise to an intensity H? The rise of the magnetic field is controlled by the self-inductance, which is $\sim 4\pi a/c^2$, and the required time T is easily seen to be given by

$$T \approx a^2 H/cE^{(e)}. \tag{4.1}$$

Putting $H = 10^{-6}$ gauss, $T = 10^9$ years, $a = 10^{23}$ cm, in (4.1) demonstrates that the battery must have an EMF of $\sim 10^{13}$ c.g.s. units, i.e. 3×10^{15} volts. This simple example shows that the sheer size of a galaxy demands, within the framework of the usual Maxwell theory, the

181

existence of an enormous impressed EMF if the astronomically observed magnetic fields are to be built up within the lifetimes of galaxies.

There have been many attempts to evade the need for an enormous impressed EMF by arguing that an initially very small seed field might have been greatly amplified by dynamo action. All such attempts are failures, however, because they do not explain the large-scale structure of the fields. To produce amplification on a large scale, the relative motions within a dynamo must have sufficient time to undergo very many 'turns', and this is not possible within the lifetime of a galaxy. To illustrate the position, suppose the seed field initially has a generally radial structure. Owing to $V(r)$ having some dependence on r, and because of the high conductivity of the ionized gas within the interstellar medium, such an initially radial field becomes wound into a spiral structure by the gas near the galactic plane, with the number of turns of a magnetic line of force between the inner and outer regions equal to the number of differential rotations between the inner and outer parts of the galaxy, and with the magnetic intensity increasing proportionately to the number of turns. Since, in the lifetime of a galaxy, there can, at most, be 10–100 turns due to differential rotation, the initial seed field can be amplified no more than a hundredfold, requiring a seed field of at least 10^{-8} gauss in order to arrive at a spirally-twisted field of intensity 10^{-6} gauss. Since explaining the origin of a field of intensity 10^{-8} gauss is essentially just as difficult as explaining one of intensity 10^{-6} gauss, no mitigation of the basic problem is really achieved by such an argument.

By an impressed field in elementary electricity, one means a departure from the standard Maxwell equations for macroscopic bodies, a departure usually from the conductivity equation $\mathbf{j} = \sigma\mathbf{E}$, which becomes modified to

$$\mathbf{j} = \sigma[\mathbf{E} + \mathbf{E}^{(e)}], \qquad (4.2)$$

where $\mathbf{j}$ is the electric current density, $\mathbf{E}$ is the electric field, and $\mathbf{E}^{(e)}$ a vector representing the impressed EMF. In the presence of magnetic monopoles, the usual Maxwell equations are changed, however, and we can think of the changes as a kind of impressed effect, something from outside the usual system. Just as $\mathbf{j}$ together with the electric charge density ρ form a 4-vector, so there is a monopole 4-vector. To distinguish the two, write $\mathbf{j}_e, \rho_e$ for the electric 4-vector and $\mathbf{j}_m, \rho_m$ for the magnetic 4-vector. Then the classical electrodynamic equations in the 3-dimensional format normally employed for the solution of practical problems are the following:

$$\text{div } \mathbf{E} = 4\pi\rho_e, \qquad \text{div } \mathbf{H} = 4\pi\rho_m, \qquad (4.3)$$

$$c \text{ curl } \mathbf{H} = 4\pi\mathbf{j}_e + \frac{\partial \mathbf{E}}{\partial t}, \qquad c \text{ curl } \mathbf{E} = -\left(4\pi\mathbf{j}_m + \frac{\partial \mathbf{H}}{\partial t}\right), \qquad (4.4)$$

$$\frac{d}{dt}\left[\frac{m_e \mathbf{v}}{\sqrt{1 - v^2/c^2}}\right] = e\left[\mathbf{E} + \frac{\mathbf{v}}{c} \times \mathbf{H}\right], \qquad (4.5)$$

$$\frac{d}{dt}\left[\frac{m_m \mathbf{v}}{\sqrt{1 - v^2/c^2}}\right] = g\left[\mathbf{H} - \frac{\mathbf{v}}{c} \times \mathbf{E}\right], \qquad (4.6)$$

where m_e is the rest mass of the charge e and m_m is the monopole rest mass, $\mathbf{v}$ being the 3-dimensional particle velocity in either case. As always, c stands for the speed of light, should one wish to work in Gaussian units.

Referring back now to the simple problem of a wire in the form of a circle of radius a, the need for a battery of enormous EMF came from the $-\partial \mathbf{H}/\partial t$ term on the right-hand side of the second of equations (4.4), which stands alone in the usual theory. The addition here of $4\pi\mathbf{j}_m$ to $\partial \mathbf{H}/\partial t$ provides the possibility of overcoming the former difficulty, although, in itself, it does not guarantee an escape from the difficulty. It is also necessary that m_m/g be very large, vastly larger than m_e/e. In effect, the maximum equivalent EMF that the monopoles can produce is given by setting $gE^{(e)}$ equal to the monopole kinetic energy $\frac{1}{2}m_m V^2$, which leads for $E^{(e)} \approx 10^{14}$ c.g.s. units to $m_m > \sim 10^{-8}$ gm, a condition well-satisfied if m_m is equal to the Planck mass, $m_{Pl} = (hc/2G)^{\frac{1}{2}}$.

It is not hard to invent a situation in which there is a lack of precise compensation between the N and S poles, and the thought arises almost inevitably that such disparities may be the primary cause of cosmic magnetic fields. The disparity ΔN_m is limited by the condition

$$\frac{\Delta N_m}{R} g^2 = \frac{\Delta N_m}{R}\left(\frac{137e}{2}\right)^2 < \sim \frac{1}{2}m_m V^2 \approx \frac{1}{2}\frac{GN_m m_m^2}{R}. \qquad (4.7)$$

If this condition were not satisfied, monopoles with the particular polarity that happened to be in excess would be repelled away from the galaxy, a situation that would continue until (4.7) was satisfied.

Condition (4.7) requires the motions of the individual monopoles to be controlled by gravity, rather than by the magnetic field arising from the disparity ΔN_m. Because monopoles are vastly more massive than electrons and protons, the electrostatic situation in which excess charge moves to the boundary of a conducting body (giving zero field in its interior) does not arise for monopoles. Monopoles at $r = R$ are compelled by gravity to fall back into the interiors of galactic halos, so that non-zero ΔN_m inevitably generates non-zero internal magnetic fields. Likewise, I rather doubt that the analogy of longitudinal electrical plasma oscillations considered by Salpeter *et al.*[10] can be taken over *mutatis mutandis* to monopoles. Provided condition (4.7) is well-satisfied, the distribution of magnetic polarity follows monopole motions which are controlled by gravity, not by the polarity distribution itself.

The magnetic field strength generated by ΔN_m would be of order

$$H_m(R) \sim \frac{\Delta N_m}{R^2} (\tfrac{137}{2}e), \tag{4.8}$$

which, in virtue of (4.7), has an upper limit of

$$H_m(R) \sim \frac{1}{2} \frac{m_m V^2}{R} \cdot (\tfrac{137}{2}e)^{-1}. \tag{4.9}$$

Again, putting $V = 200 \text{ km s}^{-1}$, $r = 50 \text{ kpc}$, and expressing m_m in terms of the Planck mass, (4.9) gives $H_m \sim 6 \times 10^{-7}(m_m/m_{Pl})$ gauss as an upper limit in the outer galactic halo.

The disparity of N and S poles within a central distance r must also satisfy a condition similar to (4.7), but with R replaced by r,

$$\frac{\Delta N_m(r)}{r} (\tfrac{137}{2}e)^2 < \sim \tfrac{1}{2}m_m V^2, \tag{4.10}$$

where $\Delta N_m(r)$ is the disparity interior to r. This leads, as above, to an upper limit of

$$H_m(r) \sim \frac{1}{2} \frac{m_m V^2}{r} (\tfrac{137}{2}e)^{-1} \tag{4.11}$$

for the field intensity of distance r. Thus, at $r = 10 \text{ kpc}$ and with $V \simeq 200 \text{ km s}^{-1}$, (4.11) gives

$$H_m \sim 3 \times 10^{-6}(m_m/m_{Pl}) \text{ gauss}. \tag{4.12}$$

The order of magnitude agreement when $m_m/m_{Pl} = 1$ of the right-hand side of (4.12) with the observed galactic field near the solar system suggests that the monopole mass may, indeed, be comparable to the Planck mass.

The field H_m is, however, the seed field generated by the monopoles. The total field of intensity H includes the dynamo effect of the rotating interstellar gas. Since the monopoles are in radial motion, the magnetic field H_m generated by the disparity of N and S monopoles would be expected to be generally radial in its structure. The presence of rotating plasma disks in spiral galaxies must appreciably modify such radially directed magnetic fields, since ionized plasma cannot cross magnetic field lines to an appreciable extent, as transversely-moving ionized gas would have to do if the magnetic lines of force remained radially-directed. In order to maintain transverse motion, more or less in circular orbits around the galactic centre, ionized gas must wind-up the field lines into a spiral structure, thereby forcing the radially-moving monopoles to cross the field lines, which the monopoles can do because of their large individual masses. The effect is to give a total field intensity H that can be appreciably larger than H_m, say by an order of magnitude, increasing H_m, say, from $\sim 10^{-7}$ gauss to $\sim 10^{-6}$ gauss, by adding a main component to H that is generated by electric currents within the plasma. These currents have a large induction effect that is cancelled by an effective EMF generated by the monopole distribution itself, as will be shown later in this section.

The monopoles tend to annihilate the component of the magnetic field generated by electric currents in the plasma, an effect first pointed out by Parker.[13] (The monopoles cannot, of course, annihilate their own contribution to the field.) However, the transversely-moving plasma constantly regenerates the transverse component of the field, with the field acting as a coupling agent whereby kinetic energy is gradually interchanged between the rotating plasma and the monopoles, a process that is maintained so long as the rotating plasma does not lose a large fraction of its angular momentum about the centre of the galaxy. It is of interest to estimate the time interval over which the plasma rotation and the transverse magnetic field component which it generates can be continued.

Each of the monopoles travelling through the plasma is acted on by a transverse magnetic acceleration of magnitude $H(137e/2)/m_m$. Hence, each monopole moving radially in-and-out through a disk-like region of the plasma of radius r acquires a transverse velocity component of magnitude $\sim (2r/V) \cdot H \cdot (137e/2)/m_m$, the first factor here being the time of travel across the region. The kinetic energy acquired in time $2r/V$ by each such monopole is, therefore,

$$\frac{1}{2}\left(\frac{2r}{V}\right)^2 \cdot H^2 \cdot (\tfrac{137}{2}e)^2/m_m, \tag{4.13}$$

and the average rate at which each monopole travelling through the plasma acquires kinetic energy is given by dividing (4.13) by $2r/V$,

$$\frac{1}{2}\left(\frac{2r}{V}\right) \cdot H^2 \cdot (\tfrac{137}{2}e)^2/m_m. \tag{4.14}$$

Hence, the rate of acquisition of kinetic energy by the monopoles passing through each unit volume of the plasma is given by multiplying (4.14) by n_m. Thus, in terms of the monopole mass density $\rho_m = n_m m_m$, the rate of energy acquisition by the monopoles is

$$\frac{1}{2}\left(\frac{2r}{V}\right) \cdot \rho_m H^2 (\tfrac{137}{2}e)^2/m_m^2 \text{ per unit volume.} \tag{4.15}$$

This acquisition of energy is at the expense of the kinetic energy of the plasma, $\frac{1}{2}\rho_p V^2$, where ρ_p is the plasma mass density. The time interval over which such a process of energy interchange can continue is given, therefore, by dividing $\frac{1}{2}\rho_p V^2$ by (4.15), viz

$$\frac{\rho_p}{\rho_m} \cdot \frac{V^3 m_m^2}{2rH^2(137e/2)^2}. \tag{4.16}$$

For a plasma of uniform density 2×10^{-24} gram cm^{-3} (equal to the mean density of the interstellar gas in the solar neighbourhood of the galaxy), and using the value $\rho_m \approx 10^{-24}$ gram cm^{-3} obtained at the end of the previous section, the required time interval is

$$\sim m_m^2 V^3/rH^2(137e/2)^2, \tag{4.17}$$

which for $H = 10^{-6}$ gauss, $V = 200$ km s^{-1}, $r = 10$ kpc, is $\sim 2 \times 10^9 (m_m/m_{Pl})^2$ years. Longer-term persistence of the coupling between interstellar gas and the monopoles would be contingent on increasing the plasma density ρ_p, plasma being then condensed into clouds, as, indeed, is largely the case in our own galaxy. Thus, for a typical cloud density $\sim 10^{-23}$ gram cm^{-3}, and for $m_m = m_{Pl}$, (4.16) gives a time scale of order 10^{10} years, the age of the galaxy.

The present considerations suggest that the interstellar medium may well be deeply affected by forces which have, hitherto, been omitted from discussion, a possibility reinforced by continuing the argument a little further. The magnetic current density generated by the transverse monopole motions is $\sim n_m \cdot (2r/V)H(137e/2)^2/m_m$, directed in circles around the galactic centre. Such a current density within a plasma disk of thickness $2t$ and radius r generates a radial electric field of magnitude

$$E \approx \frac{4\pi n_m H}{m_m c} \left(\frac{137e}{2}\right)^2 \frac{2rt}{V}, \tag{4.18}$$

which produces an electric potential difference of

$$\sim \frac{4\pi n_m H}{m_m c} \left(\frac{137e}{2}\right)^2 2 \frac{r^2 t}{V} \tag{4.19}$$

between the two faces of the plasma disk. Putting $\rho_m \simeq 10^{-24}$ gm cm^{-3}, $H = 10^{-6}$ gauss, $V = 200$ km s^{-1}, $t/r = 0.01$, $r = 10$ kpc, gives an effective EMF according to (4.19) of $\sim 10^{16}(m_{Pl}/m_m)^2$ volts, a result which for $m_m \approx m_{Pl}$ is remarkably close to the required EMF that was estimated above to be necessary if a field of intensity 10^{-6} gauss is to be built-up in an adequately short time scale – the example of the circular wire.* We see, therefore, that, provided the monopole mass m_m is large enough,† the $4\pi j_m$ term in the second of equations (4.4) can compensate for the $\partial H/\partial t$ term. The inductance effect of the latter, otherwise, destroys all hope of understanding the origin of the magnetic fields of galaxies.

This has been a somewhat lengthy interlude in the main argument, but its importance for identifying the so-called missing mass of the galactic halos in terms of magnetic monopoles justifies the extent of the discussion.

---

* Noting the difference, however, that whereas the present effective EMF between the two faces of the plasma disk generates an appropriately oriented azimuthal magnetic field in the galactic disk, the magnetic field in the example of the circular wire was directed in meridian planes through the axis of the wire. The similarity of the two cases lies in overcoming self-inductance, which is the essential problem.
† So that the time scale (4.16) is long enough.

5 Monopole annihilation

In 1931, it was not possible, of course, to consider the hadronic couplings of monopoles, such as emerge nowadays from grand unification theories. These couplings permit the annihilation of N and S monopoles with most of the large annihilation energy, $2m_m c^2$, appearing as a burst of hadrons. There has been much argument as to what the annihilation cross-section should be, with suggestions varying from a strong interaction value of order $(h/m_\pi c)^2$, m_π being the pion mass, down to a very much smaller value arising from the weak interaction. Because the astrophysical consequences of the high value are interesting, whereas there is little of potential importance in the low value, I will choose the high alternative, but, before doing so, it is relevant to consider the preprint of Kolb *et al.*[9] who claim to have proved that either the annihilation cross-section is very low indeed or monopole halos of the kind considered here do not exist.

Kolb *et al.* argue that the capture of monopoles from a galactic halo by neutron stars (taken on a time scale of $\sim 10^{10}$ years) would catalyse far too much proton decay, unless the proton annihilation cross-section is controlled by the weak interaction. About 1 per cent of the energy of proton decay would be expected to appear as X-rays, to which an observational upper limit of luminosity, $\sim 10^{31}$ erg s^{-1}, can be set. If there is a monopole halo of the kind considered here, this upper limit on the X-ray emission of neutron stars translates into a limit on proton decay, which, in turn, places a severe limit on the monopole annihilation cross-section.

It is implicit in this argument that captured monopoles accumulate within a neutron star over the whole age of the star, thereby building up a considerable monopole number which catalyse proton decay more and more as the star ages. This does not seem to me to be correct, however, because there is a process that causes monopoles to annihilate each other almost immediately after they are captured (assuming the monopoles are of nearly equal N and S types). A captured monopole enters a neutron star at a speed $\sim c/3$. In order that such a monopole does not pass through the star and move out back into the interstellar medium, its motion has to be damped by an amount at least of order V in a single passage through the star. The same damping, for example due to nucleon–monopole scattering, would operate after capture and would within a time scale of order 1 second largely destroy the motion of the monopole relative to the nucleonic material of the star, but not quite! Because of gravity, monopoles would sink gradually into the very centre of the star, like metal pellets sinking through treacle. The monopoles are drawn by gravity into a tiny zone near the centre, with their density increasing as the ratio of the volume of the star to the volume of the tiny zone, with the total annihilation rate of all the captured monopoles increasing as this ratio, and with the ratio increasing until the total annihilation rate rises to equality with the total capture rate. Thus, the energy released within the star is not determined by the catalysis of proton decay but by the annihilation of the captured monopoles, which can readily be calculated.

The rate of capture of monopoles by a neutron star of mass M and radius R is given by

$$2\pi GMRn_m/V \text{ monopoles per unit time,} \qquad (5.1)$$

a minor post-Newtonian correction being neglected. If the whole rest energy of the captured monopoles were converted at a 1 per cent efficiency into X-rays, the X-ray luminosity would, therefore, be of order

$$0.01m_m c^2 \cdot 2\pi GMn_m R/V. \qquad (5.2)$$

Putting $\rho_m = n_m m_m = 10^{-24}$ gram cm^{-3}, $M = M_\odot$, $R = 10$ km, $V = 200$ km s^{-1}, gives a luminosity of only $\sim 5 \times 10^{20}$ erg s^{-1}, less by $\sim 10^{10}$ than the value of $\sim 10^{31}$ erg s^{-1} that is considered permissible by Kolb *et al.* No difficulty would, therefore, seem to arise on this score, and so I will proceed unimpeded by this neutron-star argument.

Since there is no canonical formula for the annihilation cross-section, I will take one that is convenient for the purposes of calculations, viz $(h/m_\pi c)^2 \cdot c/v$, so that

$$\langle \sigma v \rangle \approx c \cdot (h/m_\pi c)^2. \qquad (5.3)$$

Here, v is a general encounter speed between N and S monopoles, not necessarily the galactic velocity V. In the galaxy case, however, the annihilation rate per unit volume is $n_m^2 \langle \sigma v \rangle$, which in virtue of (3.1) is

$$\left[\frac{V^2}{4\pi Gr^2 m_m} \right]^2 \langle \sigma v \rangle. \qquad (5.4)$$

The rate of energy released per unit volume due to monopole

189

annihilation is given by multiplying (5.4) by $2m_m c^2$ and the total annihilation luminosity is then obtained by integrating with respect to r. The outcome of this procedure is easily seen to be

$$L_{ann} \simeq \left[\frac{c^3}{2\pi G^2} \left(\frac{h}{m_\pi c} \right)^2 \left(\frac{2G}{hc} \right)^{\frac{1}{2}} \right] \frac{V^4}{r_{min}} \cdot \frac{m_{Pl}}{m_m}, \qquad (5.5)$$

where (5.3) has been used for $\langle \sigma v \rangle$, and r_{min} is the smallest value of r down to which the monopole density distribution (2.2) can be used. In effect, r_{min} is a measure of the precision with which the radial in-and-out motions of the monopoles are directed towards a centre, a measure of the degree of focussing of the monopole distribution.

The numerical factor in square brackets in (5.5) is about 10^{24} in c.g.s. units, so that the annihilation luminosity in these units is given by

$$L_{ann} \approx 10^{24} \frac{V^4}{r_{min}} \frac{m_{Pl}}{m_m} \text{ erg s}^{-1}, \qquad (5.6)$$

where V is to be inserted in cm s^{-1}, and r_{min} in cm. For $r_{min} \simeq 1$ kpc and V a velocity of several hundred kilometres per sec, (5.6) for $m_{Pl}/m_m \approx 1$ has an unspectacular value of 10^{32} to 10^{33} erg s^{-1}. But, if we postulate the existence of more compact monopole halos capable of generating infall speeds of order c, the situation is very much otherwise. Putting $r_{min} = 10^{18}$ cm and $V = 10^{10}$ cm s^{-1} in (5.6) gives $L_{ann} \simeq 10^{46}(m_{Pl}/m_m)$ erg s^{-1}, for example. With perhaps 1 per cent of the annihilation energy appearing as radiation, the photon luminosity would be of galactic order. Compact monopole distributions with high measures of focussing (small r_{min}) could, therefore, appear as extremely bright objects. The potential importance to astrophysics of such a physical process hardly needs emphasis.

6 The centres of galaxies

The elements of the material of the halos of galaxies would appear to have motions that are radially directed with respect to their centres. The evidence for this point of view is so strong that it makes good sense to accept it as a basis for further argument. When we ask *why* there should be such motions, the answer inevitably has to be that the halo material of a galaxy moves radially because, at one time, it was expelled from the centre. So one concludes that if the halo material consists of monopoles, then monopoles have been expelled in large numbers from the centres of galaxies.

The conundrum of the galactic halos is that their gravitational effects show them to have masses of order 10^{13} times the Sun, whereas the compact objects situated at the centres of galaxies have gravitational effects corresponding generally to masses only in the range 10^6 to 10^8 times the Sun, and how can $10^{13} \odot$ have come out of $\sim 10^7 \odot$? Certainly not from an accretion disk moving around a black hole of mass $\sim 10^7 \odot$, the currently popular theory. Although the nerves of the popular consensus may grate at the last part of the following sentence from *Astronomy and Cosmology* by James Jeans (Cambridge University Press, 1928, page 352), something of the sort would seem to be true:

The type of conjecture which presents itself, somewhat insistently, is that the centres of the (galaxies) are of the nature of singular points, at which matter is poured into our universe from some other, and entirely extraneous dimension, so that, to a denizen of our universe, they appear as points at which matter is being continually created.

The clue to a resolution of the conundrum lies in the existence of a field with negative energy density. In theoretical physics, one usually learns that, if coupled to the creation of particles of positive energy, such a field would lead to an impossible catastrophe, because the creation of particles would yield a field that was still more negative, and, thence, to still stronger particle creation, and so on to infinity. But the argument ignores gravitation. A negative energy field being gravitationally repulsive causes explosion, with an emergence of the positive energy particles, just as the observational facts indicate for the emergence of halo material from the centres of galaxies.

Although we have no description of such a process in terms of attested quantum particle physics, simple classical theory shows what the general mathematical shape of the process would have to be. The negative-energy field turns out to be best represented by a scalar, the C-field as it has often been called. Both the pure C-field terms and the interaction of the field with particles can be expressed in an action formulation,[14] from which dynamical equations can readily be obtained. The dynamical equations then lead to the following picture.

191

The C-field becomes greatly enhanced in a strong gravitational field, suggesting that a threshold for particle creation is passed only in highly collapsed objects. Such objects, instead of imploding into black holes, as would be the case if all fields had positive energy density, implode into centres of particle creation. Creation proceeds with the C-field becoming ever more negative, until, eventually, the C-field becomes strong enough for its gravitational repulsion to change the initial implosion to an explosion. Although much of the created matter from each object is probably thrown off into extragalactic space, thereby serving to 'close' the universe, once the repulsive C-field has also expanded into extragalactic space, a fraction of the created material is held together by its own self-gravitation, so forming a galactic halo. The escaping repulsive C-field has the further effect of forcing the galaxies apart from each other, thereby giving rise to the expansion of the universe.

This picture disagrees with the currently popular cosmologies in which everything of main importance to theoretical physics is over and done with in a first fleeting moment following a supposed origin of the whole universe. But the popular cosmologies lead into a morass when one tries to understand the properties of galaxies, and they also contradict the uniformity principle of James Hutton according to which the present-day world has to be comprehensible in terms of ongoing processes, a principle which should be just as applicable in cosmology as it is in geology.

Classical theory does not give information concerning the kind of particles to which the C-field is coupled. However, the arguments of earlier sections suggest that the particles are monopoles. Monopoles then become the primary material of the universe. They not only constitute the halos of galaxies but may even serve to close the universe, and it is quite possibly the clouds of hadrons resulting from monopole annihilation that provide the nucleonic component of the universe, as seen in stars, the interstellar gas, planets, and in ourselves.

It is a pleasure to thank Professor William A. Fowler for valuable discussions and correspondence on this whole subject, and especially for the remark that follows equation (3.5).

References

1 J. P. Ostriker & P. J. E. Peebles, *Astrophys. J.* **186**, 467 (1973).

2 J. P. Ostriker, P. J. E. Peebles & A. Yahil, *Astrophys. J. (Letters)* **193**, L1 (1974).
3 J. Einsato, A. Kaasik & E. Saar, *Nature* **250**, 309 (1974).
4 E. L. Turner, *Astrophys. J.* **208**, 20 and 304 (1976).
5 H. Spinrad, J. P. Ostriker, R. P. S. Stone, L-T. G. Chiu & A. G. Bruzual, *Astrophys. J.* **225**, 56 (1978).
6 C. W. Allen, *Astrophysical Quantities*, 3rd edn., p. 284, The Athlone Press, London (1973).
7 J. Kwon, F. Boehm, A. A. Hahn, H. E. Henrikson, J.-L. Vuillemier, J.-F. Cavaignac, D. H. Koang, B. Vignon, F. v. Feilitzsch & R. L. Mössbauer, *Phys. Rev.* **D24**, 1097 (1981).
8 B. Cabrera, *Phys. Rev. Lett.* **48**, 1370 (1982). Cabrera's results set an upper limit of $6 \times 10^{-10} \, cm^2 \, s^{-1} \, sr^{-1}$ for the monopole flux at the Earth.
9 E. W. Kolb, S. A. Colgate & J. A. Harvey, 'Monopole Catalysis of Nucleon Decay in Neutron Stars', Preprint, received August 11, 1982.
10 E. E. Salpeter, S. L. Shapiro & I. Wasserman, 'Constraints on Cosmic Magnetic Monopoles Imposed by the Galactic Magnetic Field', Preprint, received September 20, 1982.
11 F. Hoyle, 'What is a Spiral Galaxy', University College, Cardiff, Preprint Series No. 76, 1982, and *Astrophysics and Space Science*, in press.
12 D. Burstein, V. C. Rubin, N. Thonnard & W. K. Ford Jr., *Astrophys. J.* **253**, 70 (1982).
13 E. N. Parker, *Astrophys. J.* **160**, 383 (1970).
14 J. V. Narlikar, *General Relativity and Cosmology*, Macmillan, New York (1979).

15

The inadequacies of quantum field theory

P. A. M. Dirac
Florida State University

My whole talk will be centered around the equation

$$ih\frac{du}{dt} = uH - Hu$$

The new quantum mechanics introduced by Heisenberg in 1925 has noncommuting quantities as dynamical variables. That is to say UV and VU is not the same for two dynamical variables U and V. This was a very surprising idea, and it was hard for physicists to get used to it. I heard that Heisenberg, himself, thought his theory must be wrong when he fully realized its implications. It needed the encouragement of his professor, Max Born, to continue this way. Born was an expert of matrices, whereas Heisenberg hardly knew what a matrix was. A special interpretation of this quantum mechanics was needed, and it turns out that this interpretation has to be of statistical nature. One just calculates possibilities.

The interpretation of quantum mechanics has been dealt with by many authors, and I do not want to discuss it here. I want to deal with more fundamental things.

The equation that I have written down is the fundamental equation of motion for dynamical variables in Heisenberg's theory. The Hamiltonian H is characteristic for the dynamical system. Different choices of H belong to different dynamical systems. The development of

194

the theory soon showed that there was a close analogy between the new mechanics of Heisenberg and the old mechanics of Newton. It was later developed by several people. There was a close analogy which enabled one to find the corresponding quantum system when one had a particular classical system, and it appeared from this analogy that the quantum theory gave results very close to the old classical theory when one dealt with large masses.

All this was very satisfactory, and it leads to a great confidence in Heisenberg's theory. The theory was developed in the two or three years following 1925. Previously, we could not do better than work with Bohr's orbits. Bohr's orbits were satisfactory for simple problems where one was concerned essentially with just one electron, but, if one has two electrons interacting with each other, the Bohr theory was not at all precise. This older theory got replaced by Heisenberg's quantum mechanics, which was a general system of mechanics and people were happy to be able to use it instead of the primitive theory of Bohr orbits.

How did this quantum mechanics develop? People were very impressed by the close connection between the new theory and the old classical theory. The problem is to find the correct Hamiltonian to use for any dynamical system.

If you take a general system, such as particles and fields interacting with each other, you can handle this by classical mechanics and that suggests a certain Hamiltonian; its form has been worked out. But if this Hamiltonian is substituted into the fundamental equations of motion of the Heisenberg theory, the result is definitely wrong. It is not only wrong – it is not a sensible result at all. It is a result which has infinities in it. It is really a wrong theory, but still physicists like to use this Hamiltonian which is suggested by classical mechanics.

How then do they manage with these incorrect equations? These equations lead to infinities when one tries to solve them; these infinities ought not to be there. They remove them artificially. That means they are departing from the Heisenberg equations of motion.

People do not seem to realize that they are really departing from the original Heisenberg theory and that they confine themselves to working with this Hamiltonian, with the infinities removed from it by artificial means. Indeed, there is some justification for that because rules can be set up to remove the infinities. This is the renormalization process. It turns out that, sometimes, one gets very good agreement with experiments working with these rules. In particular, if one has charged particles interacting with the electromagnetic field, these rules of renormalization give surprisingly, excessively good agreement with experiments. Most physicists say that these working rules are, therefore, correct. I feel that is not an adequate reason. Just because the results happen to be in agreement with observation does not prove that one's theory is correct. After all, the Bohr theory was correct in simple cases. It gave very good answers, but still the Bohr theory had the wrong concepts. Correspondingly, the renormalized kind of quantum theory with which physicists are working nowadays is not justifiable by agreement with experiments under certain conditions.

In spite of this, physicists have gone a long way in developing this theory; in fact, most of the physics of elementary particles over about forty years has been along these lines. People work with a Hamiltonian which, used in a direct way, would give the wrong results, and then they supplement it with these rules for subtracting infinities. I feel that, under those conditions, you do not really have a correct mathematical theory at all. You have a set of working rules. So the quantum mechanics that most physicists are using nowadays is just a set of working rules, and not a complete dynamical theory at all. In spite of that, people have developed it in great detail.

I want to emphasize that many of these modern quantum field theories are not reliable at all, even though many people are working on them and their work sometimes gets detailed results.

I would like to point out that my insistence on the need for keeping to this Heisenberg equation previously had a big success. When one is just considering one electron interacting with the electromagnetic field, one feels that the equation of de Broglie governing the waves associated with a particle has to be connected with the theory, and people who were working with it in 1926 and 1927 were dominated by this idea. But using the de Broglie wave equation in connection with the Heisenberg theory produced inconsistencies which people did not allow to disturb them very much. But I insisted on the necessity to keep to this equation of Heisenberg, and, with this insistence, I was enabled to think of a different kind of Hamiltonian – a Hamiltonian that is not suggested at all by classical mechanics. I got away from the idea that one has to have a Hamiltonian, suggested by classical mechanics, and I got a new Hamiltonian that involved a spin variable. The spin of the electron

P. A. M. Dirac at the 1968 Orbis Scientiae at the Center for Theoretical Studies,
University of Miami.

turned out to be in agreement with observation. This was a great
success, and it showed how, by departing from ideas suggested by
classical mechanics, one could make an advance in a new direction.

I feel that, in the present situation, we should insist on the validity of
this Heisenberg equation which is the basis of the whole quantum
theory. We have got to hold onto it, whatever we do. If the equation

197

gives results which are not correct, it means that we are using the wrong Hamiltonian. The foundations of the quantum mechanics of Heisenberg, which are very sound and beautiful, should not be changed.

Can we get a better Hamiltonian? The theory of Heisenberg is more powerful than classical mechanics because its dynamical variables can be of a more general nature. One usually takes these dynamical variables to be functions of dynamical coordinates and their derivatives. Heisenberg originally formulated these equations with the dynamical variables appearing as matrices. This can be generalized by allowing more general quantities as dynamical variables. They can be any algebraic quantities such that you do not have in general commutative multiplication, but it seems that one has to retain associative multiplication. This allows dynamical variables of a more general kind that are not at all suggested by classical mechanics. It could be that the dynamical variables form the elements of some group. Modern physics is much concerned with bringing such dynamical variables into quantum theory. It could be that the dynamical variables are of a more general nature, something that has not yet been thought of by physicists.

I feel that those are the lines along which physicists should concentrate their attention, rather than working with a falsification of the Heisenberg equations. Their work should be concerned with finding the correct Hamiltonian, making use of the vast possibilities of noncommuting quantities which need not be suggested by classical mechanics. That would mean some kind of degrees of freedom occurring in a fundamental way in the equations of quantum theory. The trend followed by most physicists of keeping to ideas suggested by classical mechanics and then supplementing them by certain groups is a very restricted one. I believe that one must look for some more general kind of Hamiltonian.

Some years ago, I did think of a different kind of Hamiltonian which is in conformity with the Heisenberg equations, but all its solutions are of positive energy. This new theory has very interesting equations that follow from it, but it has not led to anything of practical importance up to the present. Still, I like to mention it as an example of the lines on which one should seek to make advance. I have spent many years searching for a Hamiltonian to bring into the theory and have not yet found it. I shall continue to work on it as long as I can, and other people, I hope, will follow along such lines.

16

Dirac and the foundation of quantum mechanics

P. T. Matthews
University of Cambridge

1 Introduction

Paul Dirac died in Florida on 20th October 1984. The whole of his professional life had been spent in Cambridge, and one of the first announcements of his death in the British press was in the Cambridge evening paper. There, he was honoured with a one-line obituary which read, 'Mr. Dirac had a degree from Cambridge University'. In the following days, the national press showed itself almost equally unappreciative of his amazing achievements. This lack of popular acclaim is in marked and surprising contrast to the esteem in which Dirac is held by his fellow physicists, who number him with Newton and Maxwell among the great theoretical physicists of all time.

In this account, I will try to summarise Dirac's contribution to the foundation of quantum mechanics in its historical context. Since he did for quantum theory what Newton had done earlier for classical mechanics, this will take us into those magic years of 1925 to 1930, when quantum mechanics emerged miraculously from the European consciousness – something that future generations will surely recognise as man's greatest achievement during the twentieth century.

To begin at the beginning, Dirac was born in Bristol on 8th August 1902. His father was Swiss, his mother English. He was educated locally, and, at the age of sixteen, he went to the Merchant Venturers College and emerged in 1921 with a B.Sc. (First Class) in Electrical Engineering, which was validated by Bristol University. He was unable to find a job, but, thanks to the wisdom of Professor Hasse, he was offered a free place to read mathematics at Bristol University. Two years later, he had qualified for a second First Class B.Sc. which was never awarded, because the University statutes do not allow for the awarding of the same degree twice to one student. This performance did enable him, however, to go up to St. John's College, Cambridge, as a Senior Student of the 1851 Exhibition, where he registered for a Ph.D. The year was 1923; Dirac was then twenty-one, and he could not have arrived in Cambridge at a more opportune moment.

The leading figures in Cambridge physics at the time were Eddington, one of the major exponents of general relativity, and Rutherford, who had transferred from Manchester to the Cavendish, succeeding J. J. Thomson, and was already launched on the programme of experiments which were to unravel the basic properties of atomic nuclei. Dirac's supervisor was R. H. Fowler – a first rate all round theoretical physicist who gave him an excellent grounding in the subject. In the following two years, Dirac published seven papers[1] in a variety of topics in statistical mechanics, relativity, astrophysics, and atomic physics, including atomic radiation. During this period, he must have become acutely aware of the stark contradictions which existed between the very well established principles of classical physics and the quantum mechanics of atoms as it was then formulated ('old' quantum mechanics) and he must have realised that the reconciliation of these two disciplines was the key physics problem of the time, transcending all others in its generality and importance.

2 Classical physics

It is convenient to summarise the situation as it then confronted him. The classical physics of basic physical laws has two main branches. The first of these, particle dynamics, established in the seventeenth century by Newton, had been worked over by the nineteenth century mathematicians and been put in a particularly elegant form, which greatly appealed to Dirac, by Hamilton. The Hamiltonian of a system of particles is the energy expressed in terms of the positions, x, and momenta, p, of the particles. For a single particle in one dimension

$$H(x, p) = \frac{p^2}{2m} + V(x), \qquad (2.1)$$

where $V(x)$ is the potential. The dynamics of the system is then given by two sets of equations:

$$\dot{x} = \frac{\partial H}{\partial p}; \qquad (2.2)$$

$$\dot{p} = \frac{-\partial H}{\partial x}. \qquad (2.3)$$

Applying these general equations to the simple system (2.1), we see that (2.2) implies

$$\dot{x} = \frac{p}{m}, \qquad (2.4)$$

and (2.3) gives

$$\dot{p} = -\frac{\partial V}{\partial x}. \qquad (2.5)$$

Given the Hamiltonian of any physical system, these equations can be solved to give the orbits, $x(t)$, of the particles of the system in terms of a set of initial conditions. (If the forces between the particles arise entirely from the gravitational attractions between the masses, these can be incorporated by specifying the appropriate gravitational potentials, but we will not be concerned with gravity any further here.)

The other main branch of classical physics is Maxwell's theory, describing electromagnetic phenomena in terms of electric and magnetic fields, or equivalently of a scalar potential $\phi(x, t)$ and a vector potential $A(x, t)$. If the system does not contain charged particles, it can also be defined by a Hamiltonian, which, in this case, is a function of the potentials (and their derivatives),

$$H(\phi, A). \qquad (2.6)$$

The Hamiltonian formalism applied to this Hamiltonian reveals the propagation of waves, transmitting energy and momentum (but no mass), through empty space with constant velocity c. These waves had been identified by Maxwell with radiation, of which visible light is a particularly important example.

The particle system described by (2.1) and the radiation-wave system described by (2.6) couple if the particles carry electric charge, e. This coupling is given by the Lorentz force which can be put in Hamiltonian form by a prescription which greatly appealed to Dirac for its elegance and simplicity, and which he had already exploited in his early publications. In (2.1), the mechanical potential must be replaced by

$$V(x) \to V(x) + e\phi(x), \tag{2.7}$$

and the momentum by

$$\mathbf{p} \to \mathbf{p} - \frac{e}{c}\mathbf{A}. \tag{2.8}$$

Applying this to the kinetic energy term in the simple one particle system given above in (2.1); then

$$\frac{p^2}{2m} \to \frac{1}{2m}\left(\mathbf{p} - \frac{e}{c}\mathbf{A}\right)^2 = \frac{p^2}{2m} - \frac{e}{c}\dot{\mathbf{x}} \cdot \mathbf{A} + O(e^2),$$

giving rise to an additional interaction term

$$H_i = -\frac{e}{c}\dot{\mathbf{x}} \cdot \mathbf{A}. \tag{2.9}$$

This formula plays a key role in the subsequent discussion.

If the coupled system is dominated by externally applied fields, then Hamilton's equations (2.2) and (2.3) applied to the extended Hamiltonian obtained by combining (2.1), (2.6) and (2.9), determine the orbits of the particles, and electromagnetic energy fed into the system through the fields is transferred to the kinetic and potential energy of the particles in the orbits. If these orbits involve accelerations – in particular, if they involve oscillations – or if the particles are made to oscillate through mechanical potentials – these oscillations couple through (2.9) to the fields and generate electromagnetic radiation. In this way, mechanical energy can flow back from the particles into the radiation waves.

By the turn of the century, this very compact set of equations was supported by an enormous body of experimental evidence and practical application. In particular, it was known to work for free electrons. In 1906, J. J. Thomson, the Professor of Physics in Cambridge, was awarded the Nobel prize for establishing the existence of the electron by showing that beams of such particles, when extracted from atoms in cathode ray tubes, could be manipulated by externally applied electric and magnetic fields in precise accord with the above formalism.

3 Old quantum mechanics

By the time Thomson was awarded his Nobel prize, it had already been established that the formalism did not give the right answers for the exchange of energy between radiation and atoms. If a body is made to radiate simply by heating it up, the relation between the angular frequency ω and the intensity of the resulting (black body) radiation is completely wrong unless it is assumed, as postulated in 1900 by Planck[2], that the energy exchange takes place in discrete quanta of magnitude $\hbar\omega$ where $\hbar$ is a universal constant ($\hbar = h/2\pi$ where h is Planck's constant).

$$E_n = n\hbar\omega. \tag{3.1}$$

Planck, himself, attributed this quantisation of energy to the oscillations in the matter system, so amplitudes, x_n, of oscillators associated with frequency ω are given by the relation

$$\tfrac{1}{2}mx_n^2\omega^2 = n\hbar\omega. \tag{3.2}$$

This implies

$$\tfrac{1}{2}mx^2\omega = \hbar, \tag{3.3}$$

which determines the amplitude, x, of the lowest allowed energy level.

Five years later, Einstein[3], from his study of the photo-electric effect, postulated that Planck's quantisation should also be applied directly to the radiation and that radiation of frequency ω should behave as an assembly of massless particles – photons – with energy and momentum given by

$$E = \hbar\omega, \qquad \mathbf{p} = \hbar\mathbf{k}, \tag{3.4}$$

where k is related to the wave length λ,

$$k = 2\pi/\lambda. \tag{3.5}$$

This explained the observed phenomena but was in direct contradiction with Maxwell's wave theory of light.

The next major step was in 1909, when Thomson's students, Geiger and Marsden[4], performed their celebrated experiments showing very large deflections up to 150° of energetic α-particles scattered by atomic nuclei. These were interpreted two years later by Rutherford[5], to imply a 'solar system' model of the atom with electrons in orbits of diameter about 10^{-10} m circling about a tiny compact nucleus with radius some ten thousand times smaller.

Although Thomson had no satisfactory alternative explanation, he was extremely well aware of the total inconsistency of the Rutherford atom. It involved rapidly oscillating electrons, which, through the interaction term (2.9), should be an immediate and intense source of radiation. A simple calculation shows that, if the electrons are set going in Coulomb orbits of the required size, in a matter of 10^{-11} seconds, they radiate away all their energy in a brief flash of light and collapse into the nucleus! No one knew this better than Thomson, and he was somewhat embarrassed by a young Danish post-doctoral student Niels Bohr, who was visiting Cambridge at the time and persisted in taking the Rutherford atom seriously. In 1912, Bohr was encouraged by Thomson to join Rutherford in Manchester, where he could pursue his ideas without bringing the University into disrepute. Bohr, under Thomson's influence, was concerned with the stability of the Rutherford atom, but amazingly was not aware of the by then well established experimental fact that, when an atom does radiate, it is by discrete spectral lines and not in a continuous flash, as implied by Maxwell's theory. Fortunately, Bohr made a brief trip to Copenhagen where he learnt, almost by chance, of the Balmer Series[6]. Armed with this information, he was led very quickly to his famous rules[7]. Planck's constant has the dimensions of angular momentum. The obvious generalisation of the Planck–Einstein quantisation rule to the Rutherford atom was that the angular momentum of allowed orbits (taken circular for convenience) should be an integer multiple of $\hbar$. Expressed in terms of the radius r and the angular frequency ω, this implies, for the ground state,

$$mr^2\omega = \hbar, \qquad (3.6)$$

which is very similar to Planck's rule (3.3). Combined with the Coulomb potential in (2.1) through (2.2) and (2.3), this determines an orbit for the ground state of hydrogen with radius

$$r = \hbar^2/me^2, \qquad (3.7)$$

and a discrete set of allowed energy levels

$$E_n = -\frac{1}{2}\frac{e^2}{r}\frac{1}{n^2} \qquad (n = 1, 2, \ldots). \qquad (3.8)$$

Bohr got rid of the radiation predicted by Maxwell by fiat. He simply asserted that the interaction term (2.8), which had to be there, did not operate when the electron was in an allowed orbit. Instead, he postulated that the radiation occurs when the electron makes discontinuous jumps from one allowed orbit to another with the emission or absorption of a photon of frequency ω given by

$$|E_n - E_k| = \hbar\omega_{nk}. \qquad (3.9)$$

As is well known, the Bohr rules were triumphantly successful. The size of the atomic system given by (3.7) is correct. The combination of (3.8) with simple values for n reproduces the Balmer, Lyman, and other simple spectral series.

The price paid for this success was to make the clash between classical physics and (the then new!) 'old' quantum mechanics even worse. It violated Maxwell by the arbitrary suspension of the radiation which should be generated by the oscillating electron. It also flatly contradicted Newtonian mechanics, which predicts that charged particles (electrons) moving in smooth electric and magnetic fields (and nothing is smoother than a Coulomb potential) should follow smooth orbits and *not* make sudden discontinuous jumps.

It was twelve years later, in 1925, that Dirac completed his apprenticeship as a research student in Cambridge. During all this time, no significant progress appeared to be made in the reconciliation of the new quantum rules with classical theory. Instead, there developed a lamentable hodge-podge of hypotheses, principles, theorems, and computational recipes! The most ingenious cooks for these recipes, in their respective centres, were Bohr, back in his native Copenhagen, and Sommerfeld in the ancient University of Göttingen.

P. T. Matthews

4 Heisenberg's matrix mechanics

The first signs of a breakthrough came from Werner Heisenberg[8], a contemporary of Dirac's, who was then a research student in Göttingen. Heisenberg had worked with Kramers on the theory of atomic radiation, and he looked for some way of establishing a relationship between Bohr's radiation rule (3.9) and the interaction term (2.9) which must be its source. The driving term is the $\dot{x}$ of the radiating electron in its orbit. Classically, the orbit can be expressed as a Fourier series

$$x(t) = \sum_n x_n \, e^{in\omega t}. \tag{4.1}$$

To introduce some quantity which could conceivably give rise to the sort of expressions occurring in Bohr's radiation rule, Heisenberg made the inspired guess that, in an atomic situation, the orbit (4.1) is replaced by a two-index symbol

$$\mathbf{x}(t) \approx x_{nk} \, e^{i\omega_{nk}t}. \tag{4.2}$$

He further assumed, in line with (3.9), that the frequency was of the form

$$\omega_{nk} = \Omega_n - \Omega_k, \tag{4.3}$$

which immediately implies that

$$\omega_{nk} = -\omega_{kn} \tag{4.4}$$

and

$$\omega_{nn} = 0. \tag{4.5}$$

To introduce Planck's constant, Heisenberg postulated that, if x is the position of a particle of mass m, then

$$2m \sum_k x_{nk} x_{kn} \omega_{kn} = \hbar, \tag{4.6}$$

which is similar, apart from numerical constants, to (3.3) and (3.6).

For consistency, Heisenberg found that he must define a multiplication rule

$$(x^2)_{nk} = \sum_m x_{nm} x_{mk}, \tag{4.7}$$

which, to his intense dismay, implies that, in general,

$$\mathbf{xy} \neq \mathbf{yx}. \tag{4.8}$$

206

Heisenberg was extremely unsure of his argument, but, while on holiday in Heligoland in June, recovering from hay-fever, he derived from his new formalism the correct expression for the energy levels of a quantum oscillator, modifying Planck's earlier conjecture to

$$E_n = (n + \tfrac{1}{2})\hbar\omega. \tag{4.9}$$

On his return to Göttingen, he wrote up the work and sent it to Pauli, who encouraged him to show it to Max Born. Born was immediately impressed and, on 29th July 1925, submitted it for publication in *Zeitschrift für Physik*.

When Heisenberg wrote his paper, he did not know what a matrix was, and Born stared at the new product rule, (4.7), for over a week before he realised that Heisenberg had re-invented matrix multiplication. He immediately set about trying to recruit a research associate with the appropriate mathematical background, but the new mathematics was so unfamiliar to physicists that he had considerable difficulty in finding anyone for the post. Fermi passed briefly through Göttingen at the time but left earlier than he had intended, as he did not like what seemed to him a philosophical and unphysical approach. Pauli turned the job down, though he immediately started work on the new theory. In the end, Born made a train journey from Göttingen to Hamburg on which he described to a colleague the new ideas and his difficulty in recruiting an assistant. By chance, the conversation was overheard by Paul Jordan, then working with Courant on Courant and Hilbert's *Methoden der Mathematische Physik* (1924), the first chapter of which is devoted to matrix algebra. Jordan introduced himself at Hamburg station and, thus, began the extremely fruitful collaboration of Born, Heisenberg, and Jordan, subsequently known as the 'drei Manne Arbeit'. But it is time to get back to Dirac.

5 Dirac's first breakthrough

On 25th July 1925, Heisenberg delivered a lecture at the Kapitza Club in Cambridge. He did not mention his new work but talked on 'Term-zoology and Zeeman botany', describing his previous collaborative work with Kramers. Dirac probably attended the lecture, but there seems to have been no significant contact between the two men.

P. T. Matthews

However, in September of that year, Bohr sent a copy of the proofs of Heisenberg's paper to Fowler, who passed them on to Dirac. Dirac's first impression was that there was little in the paper, and, after a cursory reading, he put it away. Returning to it a week later, he somehow realised that the non-commutative multiplication rule, which Heisenberg regarded as the major weakness of his new ideas was, in fact, the key to the whole problem. Ruminating over this on a lone Sunday walk, he got the idea that this non-commutative algebra might be connected with the Poisson brackets of classical mechanics, but he could not remember precisely what a Poisson bracket was. He hurried back to Cambridge, but, being Sunday, all the libraries were locked, and it was not until the following morning that he could look up the definition and everything fell into place.

The Poisson bracket is defined, for a single particle,

$$\{u,v\} \equiv \frac{\partial u}{\partial x}\frac{\partial v}{\partial p} - \frac{\partial v}{\partial x}\frac{\partial u}{\partial p}. \tag{5.1}$$

A number of identities follow directly from the definition, for example

$$\{u,v\} = -\{v,u\}, \tag{5.2}$$

and most importantly, the Jacobi identity

$$\sum_{\text{cyc}}\{u,\{v,w\}\} = 0. \tag{5.3}$$

It follows trivially from the definition that

$$\{x,x\} = 0 = \{p,p\}, \tag{5.4}$$

and that

$$\{x,p\} = 1. \tag{5.5}$$

Using the definition (5.1) and Hamilton's equations (2.2) and (2.3), it is very simple to show that

$$\dot{x} = \{x,H\}, \qquad \dot{p} = \{p,H\}.$$

Thus, for any function of position and momentum, $F(x,p)$, Hamilton's equation of motion is

$$\dot{F} = \{F,H\}. \tag{5.6}$$

In this way, the Poisson brackets provide another elegant and general

Dirac and the foundation of quantum mechanics

formulation of Newtonian mechanics. It is also easy to see from the definition how Dirac's subconscious threw up the Poisson bracket when he was thinking of the commutator (or Dirac bracket) of Heisenberg's quantum observables, which we define as

$$[\mathbf{u}, \mathbf{v}] \equiv \mathbf{uv} - \mathbf{vu}. \tag{5.7}$$

Dirac's first observation was that the general properties of the Poisson bracket, including the Jacobi identity, are all also satisfied by the Dirac bracket. It is, thus, algebraically consistent to replace Poisson brackets in the classical mechanics of any system by a constant multiple of the Dirac bracket. Dimensions require this multiple to be proportional to Planck's constant, and reality conditions imply that it is purely imaginary. Agreement with experiment is obtained if

$$\{u, v\} \to i\hbar[\mathbf{u}, \mathbf{v}]. \tag{5.8}$$

The two important equations (5.5) and (5.6) lead, respectively, to the quantum condition

$$[\mathbf{x}, \mathbf{p}] = i\hbar, \tag{5.9}$$

and the (Heisenberg) equation of motion

$$i\hbar\dot{F}(\mathbf{x}, \mathbf{p}) = [\mathbf{F}, \mathbf{H}]. \tag{5.10}$$

Dirac had hit upon a completely general rule for constructing the quantum mechanics of any physical system for which there was a well defined classical Hamiltonian. The two systems went smoothly into each other in the limit of $\hbar$ going to zero. All he had to do was to check the implications. Take Heisenberg's ansatz that

$$\mathbf{F}(t) = F_{nk}\, e^{i\omega_{nk}t}. \tag{5.11}$$

Then

$$p_{nk} = m\dot{x}_{nk} = imx_{nk}\omega_{nk}\, e^{i\omega_{nk}t}. \tag{5.12}$$

Evaluating a diagonal term of (5.9), for which the time dependence vanishes, implies

$$m \sum_k (x_{nk}(x_{kn}\omega_{kn}) - (x_{nk}\omega_{nk})x_{kn}) = \hbar, \tag{5.13}$$

which, thanks to (4.4), is identical with Heisenberg's condition (4.6).

209

Next, taking

$$\mathbf{F}(t) = \mathbf{H}(t), \qquad (5.14)$$

equation (5.10) shows that

$$H = 0, \qquad (5.15)$$

implying that $\mathbf{H}(t)$ is a diagonal matrix, independent of time,

$$H_{nk} = E_n \delta_{nk}. \qquad (5.16)$$

Finally, taking

$$\mathbf{F}(t) = \mathbf{x}(t), \qquad (5.17)$$

in (5.10),

$$\hbar\omega_{nk}x_{nk} = x_{nk}H_{kk} - H_{nn}x_{nk}, \qquad (5.18)$$

which reduces by (5.16) to the Bohr frequency condition (3.9). It worked!

Dirac knew that he had solved, in principle, the key problem which had baffled everyone for the previous decade. He said many years later that nothing he did subsequently ever gave him so much satisfaction as this, his first major discovery. Over the moon with excitement, Dirac showed his work to Fowler, who submitted it on 7th November 1925 to the *Proceedings of the Royal Society*[9], arranging for it to be published immediately, thus avoiding the delays of the usual refereeing procedure. In fact, 'the three men' in Göttingen independently arrived at the key equations (5.9) and (5.18), but the beautiful link with classical mechanics through the Poisson brackets was undisputedly Dirac's very own contribution, which established him on the international scene.

Dirac went feverishly to work and submitted another paper[10] in January 1926, in which he indicated how the Bohr levels could be obtained using his non-commutative algebra – a calculation that was completed by Pauli. In March, yet another manuscript was ready[11], in which he developed the quantum theory of angular momentum and applied it to the Zeeman effect giving energy level splitting in a weak magnetic field. In May, Mr. Dirac was awarded a Ph.D. This was the 'degree from Cambridge University', correctly reported in the one-line obituary more than fifty years later.

6 Schrödinger's wave mechanics

Dirac's triumph was far from complete. The mathematics which he was using was unfamiliar to physicists, and the matrix mechanics which he and the Göttingen school were developing seemed, to many, abstract and unphysical. In other parts of Europe, a completely different approach was making progress, which got a much more enthusiastic reception.

This began with the work of Louis de Broglie[12], in Paris, who, in 1923, had suggested that Einstein's relations (3.4), which determined the particle properties, E and p of light waves, could be read backwards to define wave properties ω and λ of particles, particularly electrons. The idea appealed greatly to Einstein, who pressed it at the Solvay Conference in 1924. It was agreed that experimental proof of electron waves would follow from the observation of diffraction patterns of electron beams deflected by crystals. James Franck, who had been the head of the department of experimental physics in Göttingen, maintained that such experiments had already been done by Clinton Davisson at the research laboratories of the American Telephone and Telegraph Company in New York. (The same laboratories which subsequently discovered the 4° cosmic background radiation.) Davisson was not convinced of his own results at the time, and they were not actually published until three years later, after the experiments had been repeated in collaboration with Germer[13].

The Solvay Conference was attended by Debye, who had retired from the Chair in Zurich in 1920 and been replaced by Erwin Schrödinger. Schrödinger was fifteen years older than Dirac and Heisenberg who had previously been a student and lecturer in Vienna. On his return to Zurich, Debye asked Schrödinger to give a seminar on the ideas of de Broglie, which he had heard about at the conference. Schrödinger did rather more than that.

His intense interest was aroused and he was struck by the apparently rather trivial observation that Bohr's quantum condition for a hydrogen-like atom, (3.6) which can be written

$$pr = n\hbar, \tag{6.1}$$

when combined with de Broglie's interpretation of (3.4) and (3.5) for the electron waves, yields

$$2\pi r = n\lambda. \tag{6.2}$$

The circumferences of the allowed circular Bohr orbits are integer multiples of the electron's de Broglie wave length. This exact, though somewhat quixotic, fitting of the waves to the available space put Schrödinger in mind of an eigenvalue equation. On 27th January 1926, he submitted a clear, self confident, brilliant paper[14]. If the momentum in the Hamiltonian for the hydrogen atom is replaced by the derivative,

$$p \to -i\hbar\frac{\partial}{\partial x}, \tag{6.3}$$

then the eigenvalue equation of the resulting operator – what we now call the Schrödinger Equation – is:

$$H\left(x, -i\hbar\frac{\partial}{\partial x}\right)u_n(x) = E_n u_n(x). \tag{6.4}$$

Schrödinger showed that the eigenvalues E_n are identical with the Bohr levels. A month later[15], he used the same technique to determine the energy levels of the harmonic oscillator (reproducing Heisenberg's earlier result) and of the rigid rotator, which he used as a model for a diatomic molecule. Of Heisenberg's work, he wrote, 'In der Methode ist er so *toto genere* verschieden dass es mir bisher nicht gelungen ist, der Verbindungsglied zu finden' (His methods are so totally different that I have not been able to see the connection) but added, 'Es schwer wird, daran zu zweifeln, er enthaltet, jedenfalls einen Teil der Wahrheit' (Nevertheless, it is hard to doubt that they contain an element of the truth). But another month later, he had solved this problem, too, in a third monumental paper[16].

He observed that, if one takes position, x, to be an ordinary variable and uses (6.3) for momentum, one has a simple operator representation for the Dirac/Heisenberg quantum condition (5.9). Given any physical observable, F (e.g., position, energy, angular momentum), expressed in Hamiltonian form $F(x, p)$, one can construct the Schrödinger operator

$$F\left(x, -i\hbar\frac{\partial}{\partial x}\right). \tag{6.5}$$

From the eigenstates of the Schrödinger Equation, one can further construct matrices, defined by the relation

$$F_{mn} = \int u_m^*(x) F\left(x, -i\hbar \frac{\partial}{\partial x}\right) u_n(x)\, dx. \qquad (6.6)$$

Schrödinger showed that the matrices x_{mn}, p_{mn} and H_{mn}, so formed, were identical with those introduced by Heisenberg. He further postulated the Schrödinger Equation of motion, that for any wave function $\psi(x)$,

$$H\psi(x, t) = i\hbar \frac{\partial}{\partial t}\psi(x, t). \qquad (6.7)$$

This is a natural extension to energy of the momentum relation (6.3). The time dependent eigenfunctions, which satisfy both (6.4) and (6.7), are

$$u_n(x, t) = u_n(x)\, e^{iE_n t/\hbar}, \qquad (6.8)$$

and matrices constructed through (6.6) with time dependent eigenstates include the exponential factors postulated by Heisenberg

$$F_{mn}(t) = F_{mn}\, e^{i(E_m - E_n)t/\hbar}. \qquad (6.9)$$

In June 1926, the picture was completed by Born[17], who gave the correct probabilistic interpretation of the wave-function (ψ was replaced by $|\psi|^2$ in the galley proofs of the paper)

$$P_\psi(x) = |\psi(x)|^2. \qquad (6.10)$$

The square modulus of the wave function determines the probability of distribution of the particle in space. By disposing of the orbit, this immediately explains why an electron in an allowed Bohr energy level does not radiate. Its probability distribution, and, hence, its charge distribution,

$$|u_n(x, t)|^2, \qquad (6.11)$$

is static and should not radiate, even according to Maxwell.

Schrödinger's work was immediately and enthusiastically accepted by the physics community. Physicists were back on firm and familiar ground. Schrödinger talked the language to which they were accustomed. Everyone knew how to handle a differential equation and physics was 'physical' again.

All this happened in the few months while Dirac was cleaning up on his breakthrough, using Poisson brackets as the basis for the Heisenberg matrix formulation. A lesser man might have been discouraged, but Dirac was in full creative flight and nothing could stop him.

7 Dirac's second breakthrough

Dirac submitted for publication again in August 1926, with another astounding paper[18]. He had immediately started using the Schrödinger formalism and, quite casually, showed how Pauli's exclusion principle followed naturally from the indistinguishability of two electrons, expressed through the anti-symmetry of the two-particle wave function. The consequent statistics, but not the symmetry argument, had been found previously by Fermi[19] – hence, the name Fermi–Dirac statistics.

In the same paper, Dirac developed time-dependent perturbation theory and applied it to the interaction Hamiltonian, (2.9), to develop a theory of atomic radiation, treating the atom quantum mechanically but taking the radiation vector potential as an ordinary classical field. He was able to show that external radiation of the appropriate frequency induced absorption and stimulated emission in accordance with the Bohr formula, (3.9), and established the equality of Einstein's stimulated emission and absorption coefficients,

$$B_{mn} = B_{nm}. \tag{7.1}$$

There was still no direct explanation of the spontaneous emission of an excited atom when no external radiation was present.

Dirac's next paper[20], submitted in December 1926, was eventually to turn out to be his greatest. In it, he went way beyond Schrödinger and established the completely general mathematical framework in which quantum mechanics has ever since been formulated. A major component was his introduction of the δ-function. The structure, though not the essential content of the formalism, was greatly clarified nearly twenty years later by his bra–ket notation, and it seems sensible to summarise, taking advantage of this later development.

Observables of quantum systems are represented by Hermitian operators α in Hilbert space, the states by generalised vectors $|\psi\rangle$. The eigenstates of an observable are specified by the eigenvalue α', so the eigenvalue equation is

$$\alpha|\alpha'\rangle = \alpha'|\alpha'\rangle.\tag{7.2}$$

The eigenstates form an orthonormal set, so

$$\langle\alpha'|\alpha''\rangle = \delta(\alpha',\alpha''),\tag{7.3}$$

where the right hand side is taken as a Kronecker δ-function or a Dirac δ-function when α' is, respectively, a discrete or continuous variable. The eigenstates of $\alpha\psi$ observable also form a complete set, which leads to the very powerful expression for the unit operator,

$$1 = \underset{\alpha'}{S}\cdots|\alpha'\rangle\langle\alpha'|\cdots,\tag{7.4}$$

where, again, $\underset{\alpha'}{S}$ is a sum if α' is discrete and an integral over α' if the variable is continuous. This implies, for example, that any state $|\psi\rangle$ can be expanded in terms of the states $|\alpha'\rangle$

$$|\psi\rangle = \underset{\alpha'}{S}|\alpha'\rangle\langle\alpha'|\psi\rangle,\tag{7.5}$$

the scalar product $\langle\alpha'|\psi\rangle$ being a component of the vector $|\psi\rangle$ in the α-representation.

The states $\langle\alpha'|$ can also be used to label the elements of a matrix representation of any other observable β,

$$\beta_{\alpha'\alpha''} = \langle\alpha'|\beta|\alpha''\rangle.\tag{7.6}$$

The expression $\langle\beta'|\alpha'\rangle$ may be interpreted as the β' component of the eigenvector $|\alpha'\rangle$ in the β-representation – or vice versa. However, this expression is also the unitary transformation from the β- to the α-representation, since, by a double use of (7.4) (emphasised by the underlining),

$$\langle\beta'|\gamma|\beta''\rangle = \underset{\alpha'}{S}\,\underset{\alpha''}{S}\,\langle\beta'|\underline{\alpha'}\rangle\langle\underline{\alpha'}|\gamma|\underline{\alpha''}\rangle\langle\underline{\alpha''}|\beta''\rangle.\tag{7.7}$$

Using (7.2) and (7.3), it follows that

$$\langle\alpha'|\alpha|\alpha''\rangle = \alpha'\delta(\alpha',\alpha''),\tag{7.8}$$

so that, any observable is represented by a diagonal matrix in its own representation. If α is continuous and

$$\alpha\beta - \beta\alpha = i\hbar,\tag{7.9}$$

215

then

$$\langle \alpha' | \boldsymbol{\beta} | \alpha'' \rangle = -i\hbar \frac{\partial}{\partial \alpha} \delta(\alpha' - \alpha''), \tag{7.10}$$

so that

$$\langle \alpha' | F(\boldsymbol{\alpha}, \boldsymbol{\beta}) | \psi \rangle = F\left(\alpha', -i\hbar \frac{\partial}{\partial \alpha''} \right) \langle \alpha' | \psi \rangle. \tag{7.11}$$

In this way, differential operators are incorporated within a framework which is based primarily on the notion of matrices.

The physical interpretation of all this mathematical formalism follows from two simple rules:

(i) the possible results of the measurement of an observable α are the eigenvalues α', and

(ii) the average value of repeated observations of an observable α for a set of systems in a general state $|\psi\rangle$ is

$$\bar{\alpha}_\psi \equiv \langle \psi | \boldsymbol{\alpha} | \psi \rangle. \tag{7.12}$$

Using (7.2) and (7.5),

$$\langle \psi | \boldsymbol{\alpha} | \psi \rangle = \sum_{\alpha'} \langle \psi | \boldsymbol{\alpha} | \alpha' \rangle \langle \alpha' | \psi \rangle$$
$$= \sum_{\alpha'} \alpha' | \langle \alpha' | \psi \rangle |^2. \tag{7.13}$$

Combining this with (7.12) shows that the probability that single measurement of α, for a system in the state $|\psi\rangle$ gives the result α' is

$$P_\psi(\alpha') = |\langle \alpha' | \psi \rangle |^2. \tag{7.14}$$

Dirac already appreciated that if two observables, such as position and momentum, do not commute this implied limitation on the extent to which their numerical values could be simultaneously specified. This was later developed into the Uncertainty Principle by Heisenberg[21].

Different representations for the observables can be developed as a matter of convenience by deciding which observables to diagonalise. (Only commuting observables can be diagonalised simultaneously.) The Schrödinger representation is obtained by diagonalising the particle positions,

$$\boldsymbol{\alpha} = \mathbf{x}, \tag{7.15}$$

in which case

$$\langle x|\psi \rangle = \psi(x) \qquad (7.16)$$

is the Schrödinger wave function, and Born's physical interpretation of the wave function, (6.10), is just a special case of (7.14). Heisenberg's matrix representations of the observables arise from the diagonalisation of the energy

$$\alpha = \mathbf{H}, \qquad (7.17)$$

and Schrödinger's construction for the Heisenberg matrices from his own formulation given in (6.6) is just a special case of (7.7) with

$$\beta = \mathbf{H}, \qquad \alpha = \mathbf{x}, \qquad \gamma = \mathbf{F}. \qquad (7.18)$$

Combined with Dirac's earlier work on the quantum limit of Poisson brackets leading to (5.9) and (5.10), this defines a general procedure for the quantum mechanical description of any simple physical system, valid when the typical action of the system (length × momentum) is comparable with Planck's constant $\hbar$. To Dirac, the distinction between the Heisenberg and Schrödinger formulations had become as trivial as the choice between cartesian and polar co-ordinates.

He had established a mastery of the new quantum physics, unmatched at that time by any of his contemporaries, and he set about applying it to two outstanding problems, in both of which the matrix formulation played a crucial role.

8 The quantisation of the radiation field

In 1927 he applied his general formalism to construct a quantum theory of the radiation field[22]. The effect was miraculous. The direct application of his techniques to the pure radiation Hamiltonian (2.6) revealed that, in quantum terms, it described an assembly of photons, exactly as Planck and Einstein had conjectured. Applying a periodic boundary condition in a box to generate a discrete set of allowed frequencies and wave numbers, the energy (and similarly the momentum) content of the electro-magnetic field could be expressed

$$\mathbf{H} = \sum_{\omega} \mathbf{N}_{\omega}(\hbar\omega); \qquad (8.1)$$

where **N** is the number operator

$$
\mathbf{N} = \begin{bmatrix} 0 & \cdot & \cdot & \cdot & \cdot \\ \cdot & 1 & \cdot & \cdot & \cdot \\ \cdot & \cdot & 2 & \cdot & \cdot \\ \cdot & \cdot & \cdot & 3 & \cdot \\ \cdot & \cdot & \cdot & \cdot & \cdot \end{bmatrix}, \quad |n\rangle = \begin{bmatrix} 0 \\ 1 \\ 0 \\ \cdot \\ \cdot \end{bmatrix} \tag{8.2}
$$

and the non-zero element in $|n\rangle$ is in the $(n+1)^{\text{th}}$ position. Then

$$
\mathbf{N}|n\rangle = n|n\rangle. \tag{8.3}
$$

An energy eigenstate is specified by giving the number of photons in each allowed state. Interactions can only be expressed by changes in these numbers.

The Fourier expansion of the vector potential $\mathbf{A}(x \cdot t)$, itself, now treated as a quantum observable, turned out to be a linear superposition of matrix operators $a_{\omega'} a_{\omega'}^{\dagger}$, where

$$
\mathbf{a} = \begin{bmatrix} 0 & \sqrt{1} & \cdot & \cdot & \cdot & \cdot \\ \cdot & 0 & \sqrt{2} & \cdot & \cdot & \cdot \\ \cdot & \cdot & 0 & \sqrt{3} & \cdot & \cdot \\ \cdot & \cdot & \cdot & \cdot & \cdot & \cdot \\ \cdot & \cdot & \cdot & \cdot & \cdot & \cdot \\ \cdot & \cdot & \cdot & \cdot & \cdot & \cdot \end{bmatrix} \tag{8.4}
$$

$$
\mathbf{a}^{\dagger} = \begin{bmatrix} 0 & \cdot & \cdot & \cdot & \cdot \\ \sqrt{1} & 0 & \cdot & \cdot & \cdot \\ \cdot & \sqrt{2} & 0 & \cdot & \cdot \\ \cdot & \cdot & \sqrt{3} & 0 & \cdot & \cdot \\ \cdot & \cdot & \cdot & \cdot & \cdot \\ \cdot & \cdot & \cdot & \cdot & \cdot \end{bmatrix} \tag{8.4}
$$

218

Dirac and the foundation of quantum mechanics

Note that these matrices are (almost) the square roots of the matrix $\mathbf{N}$,

$$\mathbf{a}\mathbf{a}^\dagger = \mathbf{N} + 1, \qquad \mathbf{a}^\dagger\mathbf{a} = \mathbf{N}, \tag{8.5}$$

so that

$$[\mathbf{a}, \mathbf{a}^\dagger] = 1. \tag{8.6}$$

They have the required property of either decreasing or increasing the number of particles in a given momentum state,

$$\mathbf{a}|n\rangle = \sqrt{n}|n-1\rangle, \qquad \mathbf{a}^\dagger|n\rangle = \sqrt{n+1}|n+1\rangle, \tag{8.7}$$

i.e., of annihilating or creating photons.

Taking the fully quantised Hamiltonian for the atom and the radiation given by (2.1) and (2.6),

$$\mathbf{H}_0 = \mathbf{H}(x, p) + \mathbf{H}(\phi, \underline{A}). \tag{8.8}$$

Dirac applied the perturbation theory he had developed to the completely quantised version of the interaction Hamiltonian (2.9). The probabilities for the emission or absorption of a photon of frequency ω are proportional to the square of the matrix elements of the appropriate terms in the Fourier expansion of the vector potential, $\mathbf{A}(x, t)$ which, from (8.7), are, respectively,

$$|\langle n + 1|\mathbf{a}_\omega^t|n\rangle|^2 = n_\omega + 1 \quad \text{(emission)}, \tag{8.9}$$

and

$$|\langle n - 1|\mathbf{a}_\omega|n\rangle|^2 = n_\omega \quad \text{(absorption)}. \tag{8.10}$$

For large n_ω, both of these are proportional to the number of photons of the required frequency. This dependence of the probability on the intensity of the radiation reproduced the earlier semi-classical results for the Einstein B-coefficients. But the quantum probability for emission, given by (8.9), does not vanish when n_ω is zero, giving rise to the previously missing spontaneous emission coefficient A_{nm} and neatly providing the mechanism for the discontinuous jumps of an undisturbed atomic electron from one allowed orbit to another, so mysteriously postulated by Bohr in 1913. To complete the picture, Dirac[23] even went on to derive Kramers–Heisenberg[24] dispersion formulae, which had provided the motivation for Heisenberg's original paper in 1925, only two years earlier.

9 The relativistic electron

During 1928, Dirac published the two papers for which he is most famous[25,26], giving the relativistic theory of the electron. By that time, Uhlenbeck and Goudsmit[27] had already introduced the electron spin to explain the doublet structure of a single electron spectra, and Pauli[28] had developed the two-component formalism in terms of his celebrated σ-matrices. Klein[29] and Gordon[30] had independently quantised, by the Schrödinger procedure, the relativistic equation

$$(p^2 - m^2c^2)\psi(x) = 0. \tag{9.1}$$

This gives no hint of spin and has the further failing that the probability-current density

$$j_\mu = \psi \frac{\partial}{\partial x_\mu} \psi - \frac{\partial \psi^*}{\partial x_\mu} \psi \tag{9.2}$$

gives a probability density which is not positive definite.

Dirac solved both these problems by the observation that (9.1) can be factorised to

$$(p_\mu \gamma^\mu + mc)(p^\nu \gamma_\nu - mc)\psi(x) = 0 \tag{9.3}$$

if

$$\gamma^\mu \gamma_\nu + \gamma^\nu \gamma_\mu = 2\delta^\mu_\nu, \tag{9.4}$$

an algebra which can be satisfied by four-by-four matrices.

For the Dirac equation

$$(p^\nu \gamma_\nu - mc)\psi(x) = 0, \tag{9.5}$$

the probability density can, consistently with Lorentz transformations, be taken to be

$$\rho(x) = |\psi(x)|^2. \tag{9.6}$$

Dirac established this with considerable care. The effect of electromagnetic fields can be simply incorporated by the relativistic generalisation of (2.8). Dirac showed that the particle described by his equation has spin $\frac{1}{2}\hbar$ and a magnetic moment

$$\mu = \frac{e\hbar}{2mc} \tag{9.7}$$

in accordance with the experimentally established properties of the electron. He also derived an approximate expression for the fine structure of the energy levels of hydrogen, a problem which was subsequently solved exactly by Darwin and Gordon. The final formula is identical in form to one found by Sommerfeld in the old quantum theory, but, since the assignment of quantum numbers is completely different, this has been described by Van Vleck as 'perhaps the most remarkable numerical coincidence in the history of physics'.

In 1930, Dirac solved the awkward problem of the negative energy states implied by his equation by his ingenious suggestion that, in the physical 'vacuum' state, these levels were all full of particles satisfying Fermi–Dirac statistics and that the observable effect would be 'holes' in this infinite sea, which would behave like positive energy particles of opposite (positive) charge. He first tried to interpret these as protons[31], but, when it was clearly established by Weyl that particles and 'holes' had to have the same mass, he very explicitly predicted[32], in 1931, 'a new kind of particle unknown to experimental physics' and wrote further, 'We may call such a particle the anti-electron'. He correctly predicted pair annihilation and pair creation by two photons. The positron was discovered by Anderson in 1932.

Actually, in 'hole theory', Dirac's amazingly confident physical intuition led him marginally astray, although he still arrived correctly at the essentials of the situation. The real answer was even simpler and lay even closer to his own previous work. In the expression for the interaction of a charged particle with a vector potential, (2.9), the product of the charge with the particle velocity is the current,

$$e\dot{\underline{x}} = \underline{j}. \tag{9.8}$$

When the scalar potential is included in a relativistic notation, the combination of (2.7) and (2.9) gives an interaction Hamiltonian in which the four-vector potential couples to the charge–current density,

$$H_i = \frac{e}{c} j^\mu A_\mu, \tag{9.9}$$

and, for particles satisfying the Dirac equation, the four-vector current is:

$$j_\mu = \bar{\psi}\gamma_\mu\psi. \tag{9.10}$$

221

If $\psi(x)$ is treated, not as a Schrödinger wave function but as an operator field, in exact analogy with Dirac's treatment of the electro-magnetic vector potential, modified appropriately to allow for the different statistics, its Fourier expansion either annihilates an electron or creates a positron. The field $\bar{\psi}(x)$ plays a similar role with the electrons and positrons reversed. The relativistic, quantum field theory interaction Hamiltonian density,

$$H_{\mathrm{i}}(x, t) = -\frac{e}{c}\, \bar{\psi}(x, t)\gamma_\mu \psi(x, t)A^\mu(x, t) \qquad (9.11)$$

then has matrix elements between states which either allow for electrons or positrons to make transitions from one energy state to another or for an electron–positron pair to be annihilated or created, in each case, energy and momentum being formally conserved by the photon which is simultaneously created or annihilated by the potential A^μ. This formulation was considerably delayed by Dirac's ingenious hole theory ideas, and the elegance of an overtly relativistic quantum electrodynamics, as described above, was not fully appreciated until nearly twenty years later, when the renormalization technique to deal with inherent infinities in the theory, was developed by Feynman[33], Schwinger[34] and Dyson[35]. Dirac had discovered all the basic ingredients required for this second great leap forward, including such technical tricks as the use of the momentum representation and the 'interaction' picture which describes the time-dependence of operators and state vectors in a manner half-way between that employed by Heisenberg and by Schrödinger.

10 Epilogue

At this point, Dirac seems to have felt the need to consolidate his position. In five short years, the horizons of physics had been immeasurably extended, and a confused and contradictory amalgam of hypotheses in atomic phenomena had been replaced by a beautiful theory which was compatible not only with classical physics, but also relativity. In this development, Dirac had played a role which, in the early stages, was important, but finally came to dominate the whole scene. After establishing his relativistic equation, he turned his attention to the first edition of his textbook[36] on quantum mechanics, which has not unreasonably been compared with Newton's *Principia*. For well over thirty years and through three editions, it was outstandingly the most profound and complete treatment of the subject, which is all the more amazing since it consists entirely of his own work. Dirac's approach of quantum mechanics is so general and so powerful that it is hard to see the trees for the wood. For many years, the subject continued to be dominated by Schrödinger's methods. Gradually, through his lectures in Cambridge and, particularly, after his introduction of the bra–ket notation in the third edition in 1947, Dirac's formulation came to be recognised for what it is, a complete and definitive statement. Of all his many achievements, in the long run, this is probably his greatest.

References

1 P. A. M. Dirac, *Proc. Camb. Phil. Soc.* **22**, 132 (1924); *Phil. Mag.* **47**, 1158 (1924); *Proc. Camb. Phil. Soc.* **22**, 432 (1924); *Proc. Roy. Soc.* **A106**, 581 (1924); *Proc. Roy. Soc.* **A107**, 725 (1925); *Mon. Not. Roy. Astron. Soc.* **85**, 825 (1925); *Proc. Camb. Phil. Soc.* **23**, 69 (1925).
2 M. Planck, *Ann. d. Phys.* **1**, 69 (1900).
3 A. Einstein, *Ann. d. Phys.* **17**, 132 (1905).
4 H. Geiger & E. Marsden, *Proc. Roy. Soc.* **A82**, 495 (1909).
5 E. Rutherford, *Phil. Mag.* **21**, 669 (1911).
6 J. J. Balmer, *Verhandlungen der Naturforschenden Gesellschaft (Basel)* **7**, 548 (1885).
7 N. Bohr, *Phil. Mag.* **26**, 476 (1913).
8 W. Heisenberg, *Zeit. f. Phys.* **33**, 879 (1925).
9 P. A. M. Dirac, *Proc. Roy. Soc.* **A109**, 642 (1925).
10 P. A. M. Dirac, *Proc. Roy. Soc.* **A110**, 561 (1925).
11 P. A. M. Dirac, *Proc. Roy. Soc.* **A111**, 281 (1925).
12 L. de Broglie, *Phil. Mag.* **47**, 446 (1924).
13 C. Davisson & L. Germer, *Nature* **119**, 558 (1927).
14 E. Schrödinger, *Ann. d. Phys.* **79**, 361 (1926).
15 E. Schrödinger, *Ann. d. Phys.* **79**, 489 (1926).
16 E. Schrödinger, *Ann. d. Phys.* **79**, 734 (1926).
17 M. Born, *Zeit. f. Phys.* **37**, 863 (1926); **38**, 802 (1926).
18 P. A. M. Dirac, *Proc. Roy. Soc.* **A112**, 661 (1926).
19 E. Fermi, *Zeit. f. Phys.* **36**, 902 (1926).
20 P. A. M. Dirac, *Proc. Roy. Soc.* **A113**, 621 (1926).

21 W. Heisenberg, *Zeit. f. Phys.* **43**, 172 (1927).
22 P. A. M. Dirac, *Proc. Roy. Soc.* **A114**, 243 (1927).
23 P. A. M. Dirac, *Proc. Roy. Soc.* **A114**, 710 (1927).
24 H. A. Kramers & W. Heisenberg, *Zeit. f. Phys.* **31**, 681 (1925).
25 P. A. M. Dirac, *Proc. Roy. Soc.* **A117**, 610 (1928).
26 P. A. M. Dirac, *Proc. Roy. Soc.* **A118**, 351 (1928).
27 G. E. Uhlenbeck & S. Goudsmit, *Naturwiss.* **13**, 953 (1925); *Nature* **117**, 264 (1926).
28 W. Pauli, *Zeit. f. Phys.* **43**, 601 (1927).
29 O. Klein, *Zeit. f. Phys.* **37**, 895 (1926).
30 W. Gordon, *Zeit. f. Phys.* **40**, 117 (1926).
31 P. A. M. Dirac, *Proc. Roy. Soc.* **A126**, 360 (1930); *Proc. Camb. Phil. Soc.* **26**, 361 (1930).
32 P. A. M. Dirac, *Proc. Roy. Soc.* **133**, 60 (1931).
33 R. P. Feynman, *Phys. Rev.* **76**, 769 (1949); **80**, 440 (1950).
34 J. Schwinger, *Phys. Rev.* **76**, 790 (1949).
35 F. J. Dyson, *Phys. Rev.* **75**, 486 (1949); **75**, 1736 (1949).
36 P. A. M. Dirac, *Quantum Mechanics*, Oxford Univ. Press. First Edition (1930). Second Edition (1935). Third Edition (1947).

General references

M. Jammer, *The Conceptual Development of Quantum Mechanics*, McGraw-Hill (1966).
A. Salam & E. Wigner, *Aspects of Quantum Theory*, Cambridge Univ. Press (1972).

3

INFLUENCED AND INSPIRED BY ASSOCIATION

17

At the feet of Dirac

J. C. Polkinghorne
University of Kent, England

Certain physicists attract a cloud of stories about themselves, both true and apocryphal. For this to happen, three things are necessary – the physicist must be a man of great distinction so that people want to know about him, he must be a person of strong and interesting character, and he must be someone who is held in affection and regard. Dirac fulfilled all these criteria, and he is the subject of much oral tradition in the physics community. All the stories centre round the way he brought to everyday life the logical directness which he employed so successfully in making his great discoveries. One example, with a local Cambridge flavour, will suffice.

The Colleges of Cambridge used to set every year highly testing examinations which were taken by able school children. Those who did well became Scholars of their College, an achievement which has been the starting point of many academic careers. As Lucasian Professor, the successor of Newton, Dirac would normally have been in a position too elevated to be concerned with such routine matters. However, in wartime, many people perform unexpected duties, and it is said that in the 1940s he was asked by his College, St. John's, to be a Scholarship examiner. He duly set an exacting set of questions. The next year, he was asked to serve again. To their surprise, his fellow examiners found that he had proposed exactly the same set of questions as the year before. On being challenged, he simply replied, 'Well, they were good questions last year and they will be good questions this year'.

I did not know any of these stories when I came up to Cambridge as an undergraduate Scholar in 1949. However, I soon heard some of them and learnt that their subject was the greatest British theoretical physicist since Clerk Maxwell. I remember the first time I saw Dirac, in the foyer of the Arts School, where the lectures for the Mathematical Tripos were given. His tall, rather gaunt, figure was one of obvious distinction. Later, I attended his lectures on quantum theory. They were based closely on his famous book, *The Principles of Quantum Mechanics*, and his audience, in addition to final year undergraduates, included a number of visiting scholars who rightly deemed it an important experience whilst in Cambridge to hear an account of quantum theory 'straight from the horse's mouth'. Attending them was one of the great intellectual experiences of my life. One was given an object lesson in how clear and elegant mathematical thinking was the key to the understanding of the structure of the physical world. We were carried along in the telling of a thrilling story of insightful enquiry. The lecturer's clarity and power of mind shone through every word he spoke, but there was absolutely no attempt to underline in any way his own very considerable contributions to the subject.

Great men usually see the one needful thing with great exactitude. They also see it the way it came to them; the intensity of their vision is concentrated on the way they first made the discovery. There is something of this idiosyncracy in Dirac's book, as there was in his lectures. Mathematically, bras and kets are a somewhat laborious (pedestrian, even) way of developing the notion of Hilbert space. Physically, Dirac was astonishingly uninterested in the great unresolved interpretative issue of the act of measurement in quantum theory; the collapse of the wave packet is simply attributed to the unanalysed concept of 'disturbance'. Nevertheless, *The Principles of Quantum Mechanics* is one of the great intellectual classics of the twentieth century and will take its place with Newton's *Principia* and Maxwell's *A Dynamical Theory of the Electromagnetic Field* in the select library of books which have formed our understanding of the physical world.

Later, when I became a research student and, eventually, a junior colleague of Dirac's, I used to see him at our weekly theoretical physics seminar. Once or twice a year, he spoke himself about his current work in progress. To young men in a hurry, thinking about pion physics and dispersion relations and all that, it seemed as though the grand old man was lost in the past, forever playing around with some ingenious version of relativistic quantum mechanics. 'All very clever', we thought to ourselves, 'but Dirac would probably not know a pion if he saw one'. The last laugh is where it ought to be, with the truly great and insightful, not tossed about by every wind of physical fashion, but profound in his understanding of the quantum field theory he had invented. I realise now, with hindsight, that I heard Dirac talk about monopoles and the quantum mechanics of constrained and of extended systems and the difficulties of quantising gravity, all topics of the highest contemporary interest, to which he contributed the unique clarity and force of his understanding.

One final story. In 1956, it was discovered that parity was not, after all, conserved in weak interactions. One of us whipper-snappers had the temerity to ask Dirac what he thought about that. With characteristic simplicity and directness, he replied, 'I never said anything about it in my book'.

18

Reminiscences of Paul Dirac

Nevill Mott
Cavendish Laboratory, Cambridge

Paul Dirac, born in August 1902, was three years my senior, and when I started research in Cambridge on theoretical physics, he was already well established. I had taken the mathematical tripos, but with the intention of becoming – if I could – a theoretical physicist, and when, in 1926, the new quantum mechanics burst upon the world, I knew at once that this was what I had to understand, and set out to do so. There were no lectures on the subject; the professors as well as the students had to study the original papers of Heisenberg, Born, and Schrödinger. In my case this entailed learning some German first. I spent the best part of a year on the job, with a German dictionary in one hand and a treatise on differential equations in the other. I also studied Dirac's work on the spontaneous emission of radiation, which gave a deduction from quantum mechanics of the Einstein A and B coefficients. As far as I can remember, Dirac was not brought in to give a much needed lecture course on quantum mechanics. Neither did I go to him for help, although we were both in the same college (St. John's). He was, by repute, rather difficult to approach.* A contemporary, sitting next to

---

* See, for instance, the comments in *Cambridge Physics in the Thirties*, p. 104, ed. John Hendry, Adam Hilger, Bristol (1984), by (Sir) Alan Wilson, on the poor relationship between theory and experiment in the Cavendish. Fowler worked mainly at home and, to see him, you had to drop in half a dozen times before you could find him in; Dirac was unapproachable and spent much of his time abroad.

him at dinner, is said to have asked him what he was working on, and got the reply – do you know what adiabatic invariants are? His interlocutor answering that he did not, Dirac said, 'What, then, is the use of my talking to you if you don't know the very elements of the subject?'

For my part, having mastered Schrödinger's equation and the probability interpretation as given in Max Born's paper 'Wellenmechanik der Stossvorgange', I set out to obtain the first wave-mechanical derivation of Rutherford's scattering law. Here, I do remember that, discussing the work with my supervisor, R. H. Fowler, Dirac was called in to give advice. After this paper was published, I was in correspondence with Charles Darwin, then Professor of Theoretical Physics at Edinburgh, who hoped that I would work with him on developing a relativistic wave equation that incorporated spin, and then could be applied to the scattering of electrons by nuclei. Darwin did, indeed, publish a paper making some attempts in this direction (*Proc. Roy. Soc.* **A116**, 227 (1927)). Then, suddenly, there appeared in the *Proceedings of the Royal Society*, Dirac's paper on the relativistic wave equation (*Proc. Roy. Soc.* **A117**, 616 (1928)), which showed that, if the electron obeyed quantum mechanics *and* the principle of relativity, it *must* have a spin. Otherwise, it would be impossible to write down an equation of first order in $\partial/\partial t$ for ψ, which was essential if the initial form of ψ, incorporating information about position and momentum, was to determine its behaviour at subsequent times. This seemed and still seems to me the most beautiful and exciting piece of pure theoretical physics that I have seen in my lifetime – comparable with Maxwell's deduction that the displacement current, and, therefore, electromagnetic waves must exist. I was immensely impressed – as were the few senior people in Cambridge who were competent to judge such things. And I also remember that, until it appeared in print, I and my contemporaries had heard nothing about it.

Then came another paper from Charles Darwin (*Proc. Roy. Soc.* **A118**, 654 (1928)). He wrote,

In a recent paper Dirac has brilliantly removed the defects previously existing in the mechanics of the electron, and has shown how the phenomena usually called the 'spinning electron' fit into place in the complete theory. He applies to the problem the method of *q*-numbers and, using non-commutative algebra,

231

exhibits the properties of the electron ... There are probably readers who will share the present writer's feelings that the methods of non-commutative algebra are harder to follow, and certainly much more difficult to invent, than are operations of types long familiar to analysis ... So the object of the present work is to take Dirac's system and treat it by the ordinary methods of wave calculus.

Darwin was able to deduce exactly the relativistic hydrogen levels by this method, and also to show that his original paper was an approximation to the truth. For my part, Darwin's formulation of the Dirac equation was the basis of much of my own work during the next few years, particularly on the relativistic formula for the scattering of electrons by nuclei, the spin polarization of the scattered beam and, with H. M. Taylor, the problem of the internal conversion of gamma rays. Here, it was necessary to use the Dirac–Darwin wave functions for the K and L electrons.

I was in Cambridge as a lecturer and fellow of Caius from 1930 to 1933, and this included the 'annus mirabilis' in the Cavendish when the neutron was discovered by Chadwick and the lithium nucleus disintegrated. In the same year, came the verification by Blackett and Occhialini of Dirac's theory of the positron – a similar brilliantly simple argument. States of negative energy must exist because the energy of an electron was $C\sqrt{\{(mc)^2 + p^2\}}$, and a square root could have negative values. The negative states must be occupied, and an unoccupied state would behave like a positive electron. Blackett was too cautious in publishing his evidence, and Anderson, in the U.S., published similar evidence slightly earlier.

Dirac was, as far as I know, never interested in applying quantum mechanics to the problems of physics and chemistry, which was the occupation of most of us. Perhaps nothing that he did later produced such an impact on physics as these two achievements, made before he was thirty.

During these three years, my wife and I got to know Dirac rather well as a personal friend. I believe we were the first to introduce him to the Cambridge theatre; but he hated anything that was not crystal clear and logical, and did not seem to like the theatre very much. In Göttingen, they used to say that he believed that there is no God and Dirac is his prophet. In a letter to my parents (8 March 1931), I wrote:

Icy weather here – as everywhere else. I went down to London yesterday in

Dirac's car – very cold. Dirac ran – very gently – into the back of a lorry and smashed a headlamp.

Dirac is rather like one's idea of Gandhi. He is quite indifferent to cold, discomfort, food, etc. We had him to supper here when we got back from the Royal Society in London. It was quite a nice little supper but I am sure he would not have minded if we had only given him porridge. He goes to Copenhagen by the North Sea route because he thinks he ought to cure himself of being sea sick. He is quite incapable of pretending to thnink anything that he did not really think. In the age of Galileo he would have been a very contented martyr.

Dirac was very popular in Russia at that time; I am sure he did not notice the discomforts of life there, and once he was invited to go climbing with some Russians on the Soviet-Chinese border. I remember how he went, dressed in the tidy black suit he always wore, to practice by climbing trees on the Gog-Magog hills outside Cambridge.

Much later, when I came back to Cambridge in 1954 as Cavendish Professor, Dirac and his wife Margit were our near neighbours and very good friends. But I do not remember any interaction between Dirac's work at the time and research in the Cavendish, though his brilliant lectures on quantum mechanics, based on his book, remained as popular as ever.

19

From relativistic quantum theory to the human brain

Harry J. Lipkin
Argonne National Laboratory

In October, 1984, we heard of the death of Professor P. A. M. Dirac, one of the giants of twentieth century physics. Shortly afterwards, a conference on 'The Impact of Science on our Lives' at the Weizmann Institute heard a lecture on the latest research into the human brain, investigating the biochemical effects of anti-depression drugs on the operation of the brain. A key ingredient in this brain research was the use of positron emission tomography (PET), a technique for looking deep into the brain with the use of radioactive atoms which emit particles called positrons. The positron, discovered a half century ago, is a very unusual particle called the antiparticle of the electron. This leads us back to Dirac, who predicted the existence of the positron a number of years before its discovery. I shall now attempt to tell the fascinating story of the positron from Dirac to brain research.

In the 1920s, Dirac played a very important role in the development of the new revolutionary quantum theory of the atom. The quantum theory explained many of the puzzling features of atomic physics, which could not be understood with the nineteenth century mechanics of Newton. But the other great revolution of the beginning of this century, Albert Einstein's theory of relativity, was not incorporated into the new quantum theory. They seemed to be very different.

The quantum theory dealt with the failure of Newtonian mechanics to describe the motion of very tiny objects the size of atoms, and provided a revolutionary new approach to atomic phenomena. Einstein dealt with the failure of Newtonian mechanics to describe the motion of objects at very high speed, like the motion of the earth around the sun and the passage of light rays near the sun's surface. Einstein provided a revolutionary new approach to motion at high speeds. All attempts to put the two great revolutions together encountered formidable difficulties. Dirac was disturbed because the quantum theory, on which he had worked so hard, was not consistent with relativity. He set himself the task of finding a way to combine quantum theory and Einstein's relativity and using his new approach to describe one of the basic particles of nature, the electron.

Today, we are all familiar with the picture of the atom, with the electron moving in an orbit around the nucleus like the planets around the sun. Electrons also move freely from one atom to another in materials called electrical conductors. This motion produces the electric currents that bring us light, telephone messages, heat from electric heaters, and power from electric motors. The electrons moving in individual atoms and moving from one atom to another in radio antennas and microwave cookers radiate the electromagnetic waves which appear to us in many different phenomena like light, radio and television signals, heat, and laser beams. The new quantum theory of the 1920s explained all these properties of the electron but completely ignored Einstein's relativity. This seemed very reasonable at the time, because all these phenomena involved only slowly moving electrons and the new effects of relativity only appeared when particles moved at very high speed, near the speed of light. For slowly moving particles, Einstein's equations of motion were very nearly the same as Newton's equations of motion.

Dirac felt that it must be possible to combine relativity with quantum theory, to obtain a new theory which would also describe the motion of very rapidly moving electrons. He developed an equation which described all known properties of the electron and incorporated the principles of both the quantum theory and Einstein's relativity. It described all the electron orbits observed experimentally in the hydrogen atom very precisely and in excellent agreement with the results of experiments, including very fine effects not previously described with the old nonrelativistic theory.

But Dirac's relativistic quantum theory of the electron had a very peculiar side effect. It also described a new family of very crazy electron orbits which seemed to make no sense at all. Furthermore, Dirac's theory implied that an electron moving in one of its regular orbits in an atom could suddenly jump into one of these new crazy orbits and, at the same time, release an enormous amount of energy. Nobody understood the meaning of these crazy orbits appearing in a beautiful new theory which was, otherwise, so sensible.

Several years later, Dirac, himself, proposed the answer to the puzzle of the crazy orbits. He postulated the existence of a new particle, the positron, which was the antiparticle of the electron. It had all the same properties as the electron, except that it had the opposite sign of electric charge. The charge of the electron is negative; the charge of the positron is positive. Dirac's equation described the orbits of both the electron and of its antiparticle the positron. The crazy orbits that had perplexed everyone suddenly became reasonable when they were interpreted as orbits of a new and different particle which had the opposite sign of electric charge from the electron.

The enormous energy released when an electron jumped from a regular orbit to a crazy orbit also had a simple and revolutionary interpretation in this new picture. Einstein's theory of relativity said that matter and energy were related; that matter could be converted into energy and energy into matter. Dirac's equation described precisely this conversion. An electron and a positron could annihilate one another, and their mass would be converted into energy. The energy of an X-ray could be converted into matter by creating an electron and a positron together. The jumping of an electron into a crazy orbit was simply the annihilation of an electron against a positron in the crazy orbit.

Dirac's revolutionary breakthrough introduced the new concept of antimatter. For every particle, there was an antiparticle which had the opposite electric charge, and a particle and an antiparticle could annihilate one another and turn their mass into energy. This was an inevitable consequence of combining Einstein's relativity with the new quantum theory. Soon afterwards, Dirac's prediction was confirmed by experiment. The positron was discovered in the cosmic rays entering the earth's atmosphere from outer space. Then, many radioactive nuclei were found which emitted positrons. Many years later, the antiparticle of the proton, the antiproton, was discovered. Today, the existence of

236

antiparticles for all particles is accepted and agrees with all experimental results.

Hundreds of radioactive nuclei which emit positrons are now known, and some of these have been found to be useful in many applications, like brain research. In the work described in the meeting at the Weizmann Institute, an atom with a radioactive nucleus was used which behaved chemically like an atom naturally attracted to a particular chemical found in the brain. When such radioactive atoms were injected into a specimen of brain tissue, they were naturally attracted to this chemical. The brain specimen was then placed in an instrument called a positron emission tomograph which detects the energy released when a radioactive atom emits a positron, computes exactly where this occurred in the specimen, and produces a picture called a PET-scan which shows the precise point where any radioactive atom and the chemical that attracted it had been. The result is a precise map of the brain showing where the particular chemical related to antidepression drugs appeared. By studying the brains of animals which had undergone different treatments, it was possible to pinpoint the effects of these drugs.

The energy which produced these pictures came from the annihilation of the positrons against the normal electrons which were present in the material. Positrons, usually, are not found in the matter we see every day, which consists of atoms built from nuclei and electrons. A positron emitted into any normal material quickly finds an electron, annihilates it, and emits the energy as two gamma rays (similar to X-rays), each with a very large energy of about one half million electron volts. The gamma rays make it easy to detect positrons, since no other known phenomenon produces two gamma rays emitted exactly at the same time in opposite directions with exactly this high energy.

Sophisticated instruments, like the positron emission tomograph, use these gamma rays to detect the annihilation of a positron against an electron in a specimen of matter and pinpoint the exact spot where the annihilation occurred. These instruments contain gamma ray detectors which convert the gamma ray energy into a short pulse of electrical energy, measure the gamma ray energy and record the exact time and place at which the gamma ray was detected. This information is fed into electronic computer circuits which search for cases where simultaneous pulses of electrical energy are observed in two gamma ray detectors on opposite sides of the specimen and check that the gamma rays have exactly the right energy. This shows that two gamma rays were emitted exactly at the same time in opposite directions from the specimen. The observation of two gamma rays emitted in exactly opposite directions and observed in two different detectors in different places provides the information that enables the computer to determine the exact location of the radioactive atoms that emitted the positrons.

This kind of information is obtainable only from radioactive atoms that emit positrons. Other types of radioactive atoms also emit radiation that can produce energy pulses in detectors, but a single pulse in one detector only tells us that radiation hit that detector. It provides no information about the direction from which the radiation came. The simultaneous pulses in two detectors produced by positron annihilation tell us also the direction. The source of the radiation must have been somewhere between these two detectors and along a straight line connecting them.

The energy of these gamma rays from positron–electron annihilation, one half million electron volts, is enormous compared with the energy of only a few electron volts released when an electron jumps from one normal orbit to another in the hydrogen atom. But it is just the mysterious energy which Dirac's original equation showed was released when an electron jumped from a normal orbit to a crazy orbit. The peculiar side effect of Dirac's theory, when properly interpreted, has now not only been confirmed by experiment, but has found uses in industry, medicine, and research, which no one could have anticipated at the time when Dirac predicted the positron.

Another side effect of Dirac's equation was its prediction of the spin and magnetic properties of the electron. The electron, like many other elementary particles, spins like a top and behaves like a tiny magnet. Dirac's theory predicted this behaviour and gave the exact and correct values of the spin and of the strength of the magnet.

In 1974, Dirac was asked to review the history of these developments at an international conference on Spin Physics at the Argonne National Laboratory near Chicago. Someone in the audience asked him if he had been disturbed when he first discovered that these crazy orbits of the electron had appeared in his equation. His answer was very interesting and instructive, shedding light not only on the workings of the mind of a great physicist, but on a general approach to frontier research into the nature of matter and energy.

Dirac said that he had set himself the goal of resolving the difficulty of reconciling the two new great revolutions of quantum theory and relativity. He had worked hard to achieve this description, and he had succeeded. 'But,' said Dirac, 'one can never hope to resolve all the difficulties at once. It is natural that, in the process of resolving some difficulties, other new difficulties arise. This simply sets the stage for the next research project, the resolution of the new difficulties. This is the way that progress is continually made in our understanding of nature.'

The prediction of the electron's spin and the strength of its magnet came as a complete surprise to Dirac. He had only wanted to combine relativity and quantum theory and had thought that it would be easiest for a particle that did not spin and had no magnet. He felt that spin and magnetism could be added later on; he only wanted to solve one problem at a time. But they came as a free bonus in his theory.

Dirac's relativistic equation of the electron solved the difficulty of bringing relativity and quantum theory together. It also solved the problem of the spin and magnetism of the electron, which Dirac had not expected. Its crazy orbits introduced a new difficulty, which was later solved by Dirac, leading us into a new world of matter and antimatter, transformation of energy into matter and matter into energy, and, eventually, to radioactive nuclei useful in brain research. And these new concepts led to many new difficulties, whose resolution have led to more difficulties.

The chain of events started by Dirac's revolutionary discovery still continues, as physicists discover more and more new particles and attempt to understand the nature of the basic building blocks of the world we live in. Alongside of these developments, the biologists and biochemists have been using the side effects of Dirac's discovery to create new tools and new techniques for investigating living organisms and understanding the human brain.

Appendix The impact of Dirac's positrons on my own career

My own scientific career began with Dirac's positrons. In 1946, when I started my graduate study at Princeton, the positron had already been discovered, and the transformation of matter into energy when a positron collided with an electron had been well established. But there had not yet been any experimental confirmation that rapidly moving positrons behaved in accordance with Dirac's equation. Everybody believed that the equation was correct; it had been tested for electrons, and the positron was the antiparticle of the electron. Still, physicists are not satisfied until every possible loophole has been checked, and theoretical predictions are directly confirmed by experiment.

One way to check Dirac's theory is to study what happens when a rapidly moving positron passes close to the nucleus of a heavy atom like platinum. The electric force between the nucleus and the positron deflects the motion of the positron from its original direction, and the amount of deflection can be measured in a laboratory experiment. Dirac's equation predicted that the amount of deflection would be different from the amount predicted by the old theory. This difference results both from the new effects of Einstein's relativity and from magnetic forces acting on the positron magnet predicted by Dirac. The British theorist N. F. Mott had investigated this process for both electrons and positrons passing close to a nucleus, using Dirac's equation, and had published exact quantitative predictions for results of experiments. This process, now called 'Mott Scattering', could be observed in the laboratory by shooting a beam of positrons at a piece of platinum and looking for positrons coming out of the platinum in different directions.

Such deflection experiments were first performed by the famous British experimenter Ernest Rutherford. He investigated the internal structure of the atom by measuring the deflection of particles called 'alpha particles' when they passed through atoms. Rutherford calculated how particles are deflected by electric forces. This process has since been called 'Rutherford Scattering', and the formula that gives the exact amount of the deflection is called the Rutherford formula.

Rutherford's experiments found very large deflections and proved that the interior of the atom was very different from a homogeneous mass like a lump of jelly, as many physicists believed at that time. Rutherford's formula showed that such large deflections could not be produced when an alpha particle passed through that kind of atom. But, if nearly all of the mass of the atom is concentrated in a tiny nucleus with electrons moving in large orbits around it, the alpha particles can easily pass through the empty space between the electrons and the nucleus, come very close to the nucleus, and be strongly deflected by the strong electric forces. This is how the modern picture of the atom was discovered.

Rutherford's experiments used slowly moving particles, where the effects of relativity were negligible. The old 'nonrelativistic theory' used by Rutherford predicted that electrons and positrons would both be deflected by the same amount when they came near a nucleus. Mott's prediction from Dirac's theory was that the effects of relativity and of the magnetic forces were very different for electrons and positrons. The 'Mott scattering formula' showed that, when electrons and positrons were moving very rapidly at speeds very close to the speed of light and very close to a platinum nucleus, the deflection of an electron could be three times stronger than the deflection of a positron under the same conditions. This had not yet been checked when I started my graduate study in 1946. In my Ph.D. thesis research project, I compared the deflections of positrons and electrons emitted from different radioactive nuclei when they passed through thin foils of platinum. The results agreed with Dirac's theory, and nobody was surprised.

When I was starting my thesis work, my thesis adviser, Prof. M. G. White, suggested that I look into a different test of Dirac's theory. An Indian theorist, H. Bhabha, had studied the process of collisions between positrons and electrons using Dirac's equation. He found that, when a positron came close to an electron, there was another process that could occur in addition to the deflection of the positron by electric and magnetic forces. The electron and positron could annihilate one another and turn their mass into the energy of gamma radiation, but, then, the gamma rays could turn back into matter again and create an electron and a positron moving in a different direction.

Prof. White suggested that I try to measure this process, which is now called 'Bhabha scattering'. The effect could be observed in the laboratory by shooting a beam of positrons at a target material containing electrons and looking for positrons coming out of the target in a different direction. Some of the positrons detected would simply have been deflected by the electromagnetic forces, but there would be an additional contribution from the annihilation of the electron–positron pair into gamma ray energy and the creation of a new electron–positron pair.

I found that the experiment was not feasible at that time. Even if I used

241

the strongest possible source of positrons available at the time, and the best possible detectors, I would not get enough positrons into my detector to give a significant effect. However, I then noted that the predictions from the Dirac theory for Mott scattering had not yet been checked and found that this experiment was feasible and was a suitable subject for a thesis.

Much has changed since those days. The best detectors available for positrons and electrons were Geiger counters. A few doors down from my laboratory, in the basement of the Princeton physics building, a young professor named Robert Hofstadter was investigating the possibility of using sodium iodide crystals as detectors for radiation. But these still required extensive development before they could be used in any experiment. The best source of positrons I could obtain came from a radioactive nucleus which was made by shooting a beam of helium nuclei into a piece of copper in a machine called a cyclotron. In collisions between the helium and copper nuclei, radioactive nuclei of the element gallium were formed which decayed by emitting high speed positrons. The positron sources for my experiment were prepared in a cyclotron in Washington, D.C., by bombarding a copper target with helium nuclei for a full day. In the evening, the radioactive target was flown to Princeton by a private plane. A radiochemist then separated out the radioactive gallium from the lump of copper and deposited it into my apparatus. By midnight, I started my experiment, and I worked night and day for several days, until my radioactive source had lost its activity. The particular nucleus I used had a nine hour 'half life', which means that it loses half its strength every nine hours. After 36 hours, it was too weak to use any more.

Today, Hofstadter's sodium iodide crystals have been developed and are widely available as radiation detectors. There are machines for producing intense beams of positrons and electrons, and collisions between very high speed electrons and positrons are studied with the aim of finding new particles. One of the detectors used with these colliding 'electron–positron' accelerators is called the 'crystal ball' and consists of tons of sodium iodide.

There was tremendous progress in the development of particle physics during Dirac's lifetime. His equation and Bhabha scattering are now well established; nobody worries about testing them any more. On the contrary, the Bhabha scattering, which was considered so interesting but undetectable in 1948, is easily detected and no longer of interest in 1987. Instead, it is always present as a 'background' in the experiments looking for other new effects of greater interest in electron–positron collisions. Sometimes, Bhabha scattering is measured as a convenient way to check whether the apparatus is working properly.

20

Dirac in 1962, weak and gravitational radiation interactions

J. Weber
University of Maryland

How Dirac obtained a Social Security Number

I returned to the Institute for Advanced Study, in Princeton, during the summer of 1962. Shortly thereafter, the chairman of the University of Maryland Physics Department, Dr. John S. Toll, telephoned. He asked me to invite Professor Dirac to visit the University of Maryland and address our physics colloquium.

I spent a day gathering courage and decided to approach the greatest physicist in the world at tea. Dirac was charming and agreed readily, saying that Mrs. Dirac wished to visit the National Gallery of Art in Washington, D.C.

I was jubilant and reported the good news to Johnny Toll. His response was unexpected. He said, 'Joe, please obtain Dirac's Social Security Number.'

I was afraid to approach Dirac a second time.

There is a theorem ascribed to Professor Thomas Gold, which states that positive human qualities are correlated. The secretaries at the Institute for Advanced Study are excellent examples. They are beautiful, and efficient.

Professor Oppenheimer's secretary was very beautiful indeed, and I always enjoyed trips to the front office. I asked her to obtain Dirac's

Social Security Number from Institute files. She informed me that Dirac was a British subject and had no Social Security Number.

I telephoned Johnny Toll to advise that no Social Security Number existed for Dirac. Johnny remarked that it would be necessary for Dirac to fill out a required form to obtain a Social Security Number. Without that number, no honorarium could be paid.

Again, I approached Oppenheimer's secretary, with a request that the Institute for Advanced Study staff obtain the required information from the files for the Social Security Number application. I was soon informed that data such as the Christian names of Dirac's parents were required, that such information was not available in the Institute for Advanced Study files.

It was clear that there was no escape – I would have to ask Dirac to fill out a United States Government form. Finally I inserted this form together with a short note, in Dirac's mailbox.

Within a few hours the form was returned, with all required data, together with a polite note thanking me for arranging the visit to the University of Maryland.

It was another success for Gold's theorem. The greatest physicist in the world was well endowed with kindness and humility.

New approaches for weak and gravitational interaction physics

Dirac devoted considerable attention to the weakest interactions in nature. He had great interest in the development of gravitational radiation antennas and gave us very strong encouragement.

A major objective of twentieth century physics, beginning with Einstein, has been the development of a field theory which describes all forces in a unified way.

The most successful theory, quantum electrodynamics, is based on considerable experimental data. In contrast, very few experiments have been carried out with the gravitational interactions, primarily because of the great weakness of the gravitational forces associated with relatively small masses, in laboratory experiments.

To provide new experimental data, it was decided to study the dynamics of gravitational fields. The classical theory of elastic solid and free mass interferometer antennas was developed at the University of Maryland, together with an experimental program, beginning in 1958.

In common with many new concepts, new features of the theory, apparatus development, and observations received intense criticism before acceptance. There were great fluctuations in financial support. Professor R. P. Feynman suggested that I explore new methods in weak interaction physics as a means for achieving stable operation of a research group. The physics was so interesting that the project became all consuming. Financial stability may never be achieved!

The weak interaction and gravitational radiation theory and experiments of the period 1960–75 had almost nothing in common, except the idea that detectors should be as large as possible.

Discussion of both kinds of interactions, from the point of view of the quantum theory of scattering, leads to important new results.

Until recently, all weak interaction experiments have given total cross sections proportional to the total number N of scatterers. We shall study these, with a view to understanding a completely new method[1] which gives total cross sections proportional to N^2.

Scattering by a two dimensional array

Suppose we have a two dimensional array of N scatterers, with spacing b in the x and y directions. Let each scatterer consist of a delta function potential U given by

$$U(\bar{r}) = B\delta(\bar{r} - \bar{r}_n) \qquad (1)$$

for a scatterer at $\bar{r}_n$. For incident particles having momentum $\bar{p}$ with de Broglie wavelength small compared with b, the total cross section σ_N is given approximately by

$$\sigma_N = \frac{p^2 B^2 N}{4\hbar^4 c^2}. \qquad (2)$$

σ_N is proportional to N. This results from the fact that the differential cross section is proportional to N^2, but the solid angle into which particles are scattered is limited by interference phenomena to $1/N$. This is a consequence of the fact that phase shifts associated with the difference of incident and scattered momenta, for scatterers at opposite ends of the array, may approach π for small scattering angles.

Scattering of neutrinos and antineutrinos by rigid crystals

There is a method for enormously increasing the solid angle. For neutrinos and antineutrinos, the nucleons will scatter. A nucleon is a particle with a wavefunction, and it may exchange momentum. We may imagine a process in which all momentum is exchanged at a single scatterer. If that scatterer is tightly coupled to other scatterers, its identity cannot be established by subsequent measurements. There are now three momenta, instead of two, contributing to phase shifts. If a nucleon is so tightly bound that it can exchange a large fraction of the incident particle momentum without having its identity determined, the solid angle may approach 4π. Under these conditions, low energy neutrinos and antineutrinos will have a total cross section

$$\sigma_N = \frac{G_w^2 E_v^2}{4\pi h^4 c^4} (0.387 N_U - 0.693 N_D)^2. \tag{3}$$

Here, G_w is the Fermi weak interaction coupling constant, E_v is the energy of incident (low energy) neutrinos or antineutrinos, N_U is the number of up quarks, and N_D is the number of down quarks.

Equation (3) is enormously greater than available cross sections for other low energy processes, during Dirac's lifetime.

New approach to gravitational antennas[2]

The 1958–63 approach which appealed very much to Dirac was based on a rigorous solution of Einstein's equations and the classical theory of elasticity. The method of Fock and Papapetrou was employed to deduce equations of motion of a gravitational antenna as a continuous elastic solid.

During the past two years, a different approach has been employed.

The elastic solid antenna is regarded as an ensemble of atoms coupled by chemical forces.

For a single quadrupole of reduced mass m and mass separation r, both the classical and quantum theory give the cross section σ as

$$\sigma = \frac{8\pi^3 Gmr^2}{c^2\lambda}. \tag{4}$$

In (4), G is Newton's constant of gravitation, c is the speed of light, and λ is the gravitational wavelength. For a solid composed of N quadrupoles, single gravitons may be exchanged anywhere. The antenna length is smaller than an acoustic wavelength – very much smaller than a gravitational wavelength. Phase shifts are, therefore, negligible over the antenna. The total cross section is, therefore,

$$\sigma_{\text{TOTAL}} = \frac{8\pi^3 Gmr^2 N^2}{c^2\lambda}. \tag{5}$$

The total mass $M = nm$ and the length $L \approx N^{1/3}r$. These relations, together with (5), give

$$\sigma_{\text{TOTAL}} = \frac{8\pi^3 GML^2 N^{1/3}}{c^2\lambda}. \tag{6}$$

The factor $N^{1/3}$ is absent in the classical theory. A deeper analysis shows that it is a consequence of the equivalence principle and the quantum theory. The cross section (6) is very much larger than the classical value.

References

1 J. Weber, *Phys. Rev. C* **31**, (4), 1468–75 (April 1985); *Foundations of Physics* **14**, (12), 1185–1209 (December 1984).
2 J. Weber, *Sir Arthur Eddington Centenary Symposium, Nagpur, India, January, 1984, Proceedings*, Vol. 3, eds. T. M. Karade & J. Weber.

21

Schrödinger's cat

Willis E. Lamb, Jr.
University of Arizona

This paper contains material I presented in a lecture in July, 1985, at a meeting of Physics Nobel Prize winners. Such a meeting is held about every third year in the city of Lindau on the Bodensee (Lake Constance) in Bavaria. I have participated in nine conferences, beginning in 1959, while Paul Dirac had attended all eleven of the sessions prior to 1985. I had hoped that he would be in Lindau for the 1985 meeting, so that I could get some more of his reaction to my ideas on the nature of quantum mechanics. Although I found in past conversations with him that he was always very friendly and kind, he would not tell me much. Still, I thought it worthwhile to try again. Unfortunately, he was called elsewhere and could not attend.

1985 is the sixtieth anniversary of the birth of quantum mechanics. Sixty years is a long time. Without quantum mechanics, physics would have had a very different and much less interesting history. Quantum mechanics was founded in 1925 by Werner Heisenberg (matrix mechanics), Erwin Schrödinger (wave mechanics), and Dirac (quantum mechanics). A year later, Max Born added the essential probability interpretation for the wave function. With Dirac's death a few months ago, all four of the founders are gone, but their work will be remembered as long as science is part of our culture. Heisenberg and Born attended a number of the meetings of the physicists in the city of Lindau. I do not

249

From left to right: Francis H. C. Crick, Lars Onsager, P. A. M. Dirac, Willis E. Lamb, Jr, January 1973.

think that Schrödinger ever participated, although, after he died, Mrs. Schrödinger visited Lindau during at least one of the meetings.

In order to illustrate some of the features of quantum mechanics, I have chosen to center the discussion around the problem of Schrödinger's Cat, which dates from 1935. Before consideration of this paradox, it is desirable to briefly mention some other developments of the subject. In 1925, Cornelius Lanczos[1] invented a quantum theory involving integral equations. He was discouraged by Wolfgang Pauli's comments. The 1927 theory of Green's function propagators by Earle Kennard[2] can be regarded as an anticipation of Richard Feynman's[3] path integral formulation of quantum mechanics. Density matrices, which play a central role in parts of the present discussion, were independently invented by Lev Landau[4] (1927), Johann von Neumann[5] (1927), and Dirac[6] (1929).

Heisenberg's uncertainty principle[7] was first introduced in a 1927 paper which mentioned a 'gamma ray microscope.' He relied on mathematical properties of Fourier transforms to derive a rather primitive form of an uncertainty relation for coordinate and momentum. Heisenberg gave no serious analysis of the operation of a gamma ray microscope making proper use of quantum mechanics. In fact, to date, no one has made such a calculation. A more general uncertainty relation was discovered in 1929 by Howard Robertson,[8] and this was further extended by Schrödinger[9] in 1930. Heisenberg's Chicago lectures[10] of 1929 expanded on his 1927 discussion of the gamma ray microscope, and included a derivation of the uncertainty relation a little less general than that given by Robertson.

Schrödinger[11] had found wave packet solutions for simple problems in quantum mechanics. He very much wanted to have wave packets stay together without spreading, so that they could be used to describe particles. We now know that nonlinear partial differential equations admit of solutions called solitons which can retain their shape. The normal dispersion effects which produce spreading of wave packets can be cancelled out by nonlinearities in the wave equation. It is amusing to learn that, in 1915, Schrödinger[12] had considered a problem of a nonlinear chain which, unknown to him, had soliton solutions. My own feeling is that the progress of quantum mechanics would have been set back by many years if Schrödinger had realized that he could have kept wave packets together by introducing nonlinear terms into his wave equation.

Quantum mechanics was extended to apply to the electromagnetic field by Dirac[13] in 1927, and, in 1929, this was followed by the quantum theory of fields paper of Heisenberg and Pauli.[14] In 1928, Dirac[15] found the relativistic wave equation for an electron which has had enormous significance for the development of physics. A reconciliation of quantum mechanics and general relativity has yet to be made.

On the birthday of quantum mechanics in 1925, I was twelve years old and in the eighth grade of school, about to begin the study of algebra. By the time of the Cat Paradox in 1935, I had learned enough quantum mechanics to start research in that subject. My thesis adviser was Robert Oppenheimer, who took his doctor's degree with Born in Göttingen in 1927. I learned a lot about quantum mechanics from Oppenheimer, and a great deal from other theoretical physicists, too numerous to name here, who worked in the subject after 1925. With all of their help, I did not learn as much as I wanted to about the physical significance of quantum mechanics. I have had to try to fill in some gaps for myself, and expressed the hope that some of the young and old students in the audience at Lindau might find my remarks helpful.

My title for the lecture, 'Schrödinger's cat', was a timely reference to a book published last year by John Gribbin[16] called *In Search of Schrödinger's Cat*. Gribbin took a Ph.D. in astrophysics from Cambridge University. I do not know if he heard any of Dirac's lectures on quantum mechanics at Cambridge. Gribbin is now a consultant for the *New Scientist* magazine and a writer of semi-popular books on science. The subtitle of his book is *Quantum Physics and Reality*, and the cover claims that it is 'A fascinating and delightful introduction to the strange world of the quantum. Absolutely essential for understanding today's world.'

I can agree with some of those rather immodest claims. Gribbin's book is well written, and quite entertaining. However, it has a number of nontrivial errors. My opinion that there are flaws in his physics seems to be supported by the two reviews of the book I have seen: one by Sir Rudolf Peierls in the *New Scientist*,[17] and another by Russell MacCormmach in *Science Digest*.[18]

In 1957, a student of John Wheeler's, Hugh Everitt,[19] introduced the

'many universes' interpretation of quantum mechanics. I have no time
to deal with such an inconvenient fantasy. As far as I know, it is not
shared by many physicists. I have to warn the reader that Gribbin rather
favors this approach. It may be more appropriate for the astrophysics of
the big bang than for problems dealing with a small part of the universe.

It is a strange fact that many famous physicists did not like the
dominant role of probability theory in quantum mechanics. Among
them were Albert Einstein, Louis de Broglie, and Schrödinger. Nils
Bohr and Einstein had many arguments on the subject and considered
many gedanken or thought experiments, including the two slit
diffraction experiment in some of its many forms. Fifty years ago,
Einstein, with Boris Podolsky and Nathan Rosen,[20] published the
famous 'EPR' paper. This deals with a system of two particles which are
initially joined together, and later allowed to separate. In that same
year, 1935, Schrödinger[21] published a three part paper, entitled 'The
Present Situation in Quantum Mechanics.' One of the shorter of its
some eighty paragraphs contained the Cat Paradox. We also owe to
Schrödinger a highly regarded 1944 book,[22] entitled *What is Life?* If
time permitted, I could make some rather critical remarks about the
quantum mechanics in this book.

I have a firm belief that none of these problems were properly treated
according to conventional quantum mechanics in the discussions of any
of these distinguished people. To save time, I will deal only with the Cat
problem, but similar arguments can be applied to all such gedanken
arrangements.

Let me first remind you of some more features of quantum mechanics
which bear on the Cat Paradox. Quantum mechanics is a generalization
of classical (nonrelativistic) mechanics. The first thing to do in the
subject is to define the system of interest. That means to introduce the
coordinates and velocities (or momenta) of the various parts of the
system. There is usually a certain degree of choice here. If the system is
taken too large, the mathematics will be too difficult. If the system is
taken too small, some important features of the problem will be
neglected. If the system is not isolated, some extra terms will have to be
included in the Schrödinger wave equation to allow for external
disturbances, or the notion of the 'system' will have to be extended to
take (a little bit) more of the universe into account. Once the system and

its Hamiltonian have been defined, one should be very suspicious if anyone tries to bring a different system or a different Hamiltonian into the discussion.

In quantum mechanics, a physical system *can* have *states* which are described by wave functions which may be complex, i.e., can contain the imaginary quantity $\sqrt{-1}$. Quantum mechanics is an essentially probabilistic theory. The absolute values squared of wave functions can be used to calculate probability distributions of various functions of the dynamical variables of the system. Wave functions are functions of the coordinates which describe the configuration of the system. They depend on time in a way described by the Schrödinger time dependent wave equation for the system. For a reasonably isolated system, there are some so-called stationary states. The corresponding probability densities are independent of the time. However, there is a much larger number of nonstationary states than stationary states.

Any observation or measurement made on the system represents a disturbance which should be taken into account in the analysis of the problem. The four founders of quantum mechanics did not tell us much about how measurements are to be made, and what influence, if any, the result of a measurement might have on the subsequent history of a system. In his 1930 book, *The Principles of Quantum Mechanics*, Dirac[23] postulated that a measurement of an observable would lead to a result equal to one of the eigenvalues of that observable, and would leave the system in a new state whose wave function was an eigenfunction of the observable corresponding to the measured eigenvalue. Johann von Neumann,[24] in his 1932 book *Mathematical Foundations of Quantum Mechanics*, expanded on this idea. He distinguished between two ways in which wave functions could change with time: (a) causally, in accordance with a time dependent Schrödinger equation, and (b) acausally, as a result of a measurement, to the 'reduced' wave function exactly as postulated by Dirac. Both of these great men left many questions unanswered.

I lectured in Lindau in 1968 on preparation of states and measurement in quantum mechanics[25] and, again,[26-28] in 1982 about my unhappiness with von Neumann's wave function reduction hypothesis. If I returned again in the 1985 lecture to the measurement problem, it was because I thought that Gribben's recent book (and some others, of less merit) seemed to be giving to the general reader a wrong impression of the nature of quantum mechanics. Also, I hoped to learn something from the reactions of a perceptive and critical audience.

A very important principle of quantum mechanics was recognized by Dirac in his 1930 book, and much used by Pauli[29] in his 1933 *Handbuch der Physik* article. This is the superposition principle. For reasons which will be apparent shortly, we denote by L and D two possible suitably normalized wave functions for the system. L and D will be taken to be orthogonal to each other, i.e., they are very 'different' functions. The superposition principle states that an unlimited number of other possible states of the system exist which are of the form called superposition states, whose wave functions are linear combination of L and D. Thus, one may have the normalized wave function S given by

$$S = aL + bD,$$

where a and b are any complex constants whose sum of squared absolute values add up to unity. If the wave functions L and D happen to be wave functions for stationary states, then S will, in general, represent a nonstationary state which is neither L or D, but something 'in between'. Consider a system which is in state S. According to the conventional interpretation of the wave function, if an experiment can be devised which asks whether the system is in state S, it will give the answer 'Yes'. For a system in the same state S, we may carry out a very different experimental procedure which is devised to ask whether the system is in state L. The result for any one measurement may be either 'Yes' or 'No'. In a series of repeated measurements for an ensemble of systems of exactly the same sort, we have been considering, each in the same state S, the probability for getting the answer 'Yes' to the L question is $|a|^2$. A similar result $|b|^2$ would be obtained for measurements which recognize the state D. At this point, we should not worry about how a system is to be put into a given state S, or how experiments are to be designed to give information about the state of a system. I gave some examples of these things in my 1968 lecture and will later discuss some other simple examples of such measurements which are relevant for the cat problem.

Imagine a box for a cat that contains a radioactive source, a detector that records the presence of radioactive particles (a Geiger counter, perhaps), a glass bottle that contains a poisonous gas such as hydrogen cyanide, and a hinged hammer held above the bottle by a thread which will be cut by a knife if (and only if) the counter clicks. A living cat in good health is placed in the box. The apparatus in the box is arranged so the detector is switched on just long enough for there to be a fifty-fifty chance of a click of the Geiger counter. If the detector records a decay, the thread is cut, the hammer falls, the bottle breaks and the hydrogen cyanide gas kills the cat. Otherwise, the cat lives. We have no way of knowing the outcome of this experiment until we open the box and look inside. What is the problem? We have heard that Pandora got into trouble when she opened a box. Did our curiosity kill the cat?

There is no problem for a classical physicist. He might say that what happens is the 'will of God' or, perhaps, that it is 'just the way the cookie crumbles'. If the classical physicist is a gambler, he has experienced excitement without calling on quantum mechanics when a card is turned over, or a ball settles down after the spin of a roulette wheel. The problem exists only for a person who knows a little quantum mechanics, and who believes that quantum mechanics is universally applicable and should be applied to the cat experiment. Such a person believes that when the box is opened the box–cat system is described by a wave function like S, which is a sum of two other wave functions, L and D. One of these, L, represents a living cat, with undecayed nucleus, unbroken bottle, . . . , etc., and the other, D, a dead cat, decayed nucleus, broken bottle, . . . , etc. The constants a and b of the superposition are each taken to have equal absolute values of $1/\sqrt{2}$, so that the probability of finding a live cat is 50%. Until the observer looks to see whether the cat is alive or dead, the wave function is a superposition, or linear combination, of the two functions. The cat may be alive, or the cat may be dead. If the observer determines that the cat lives, he may be happy to accept von Neumann's sudden change of wave function from S to L, which is the wave function corresponding to a living cat.

I now return to the application of quantum mechanics to the cat problem. First of all, I have to state some things about the subject of quantum mechanics that I omitted before. I mentioned that systems can exist in states. However, in general, this means that some process has prepared the system to be in that state. It is very unlikely that a given complicated system will have a definite wave function. The best we can do is try to describe a system by a statistical distribution or ensemble of states. The density matrix of Landau, von Neumann, and Dirac is ideally suited for describing mixtures. If the system has a wave function, one speaks of a 'pure case'. Otherwise, one has a 'statistical mixture'. A pure case is all too easily converted into a mixture by any small erratic disturbance, while a mixture can never be put back into a pure case, except by some process of selection of a member of the ensemble which happens to be in a desired state. Such a selection process involves external interactions with the system that completely wipe out the memory of the past. One is reminded of Lewis Carroll's Humpty Dumpty, who, once broken, could never be put together again.

In the case of the cat problem, the system should certainly include the atoms, molecules, and macromolecules contained in the box and all of its contents. The model should make provision for the opening of the lid and for the disturbances of the system brought about by observation of the opened box and its contents. We will have at least a million million million million degrees of freedom. No such problem can be solved analytically or numerically, except in the most trivial cases. A living cat is not an isolated system. The cat–box system is not really an isolated system, because there is an observer who will open the box. Even if it were an isolated system, it could not be assigned a wave function, but only a mixture of wave functions. Even if it could have a wave function, the observation made when the box was opened would disturb the system enormously. At the very least, it would randomize the complex phase angle of the ratio b/a of the two probability amplitudes of the D and L component wave functions for the dead and live cat. Such a mixed state is called an incoherent mixture of the two states of live cat and dead cat. Such mixtures occur in discussions which attempt to apply quantum mechanics to coin tossing.[30]

I will make a few more remarks about the cat problem. It is a messy, rather pointless problem, and in bad taste for several different reasons. (1) The contemplated treatment of the cat is inhumane. (2) The possible death of the cat serves no useful scientific purpose. (3) Cats often have nine lives. (4) A different kind of bad taste involves proposing a complicated problem when a simple one will suffice to illustrate the physical principles involved.

We can easily get into a much simpler problem, which would still leave Einstein and Schrödinger unhappy. The above expression for the wave function, $S = aL + bD$, has the *appearance* of the kind of wave function which is much considered in the so-called two state or two level problem. In the cat problem, one foolishly thought of L and D as representing pure case states of the living and dead cat. In the simpler

Willis E. Lamb, Jr.

case, the wave functions L and D are not wave functions of a complicated system, but of a system with one (two valued) degree of freedom. The theory of a number of very important problems can be reduced to the theory of a two level system. A particle with a spin of 1/2 provides one example. The one particle system may have other degrees of freedom, such as translational motion, but, in some applications, only the spin degree of freedom is important. There is a very great deal of experimental work on two level systems, and it is clear that the probabilistic interpretation of quantum mechanics works very well, indeed. The theory of nuclear magnetic resonance provides examples. The recent development of NMR tomography will quite possibly win someone a Nobel prize. The theory of masers and lasers makes heavy use of the two level problem. The Nobel prizes of Stern, Rabi, Bloch, Purcell, Kusch, Townes, Shawlow, and Bloembergen are related to the two level problem. I would not be giving a lecture at Lindau without some understanding of it. Rudolf Mössbauer's lecture on the first day of the 1985 session made numerous references to two level problems in elementary particle physics.

Let us go back to an early atomic beam experiment of Otto Stern and Walter Gerlach. They heated metallic silver in an oven and made a beam of silver atoms in a vacuum chamber. (I will add a few experimental facilities which they did not have in the twenties, but could have had now.) The atomic beam is directed along the x-axis into a magnetic field which is mostly in the z-direction and which increases in magnitude as z goes from negative to positive. In the inhomogeneous magnetic field, the beam of atoms is split into two beams, one moving slightly up in z and one slightly down. Quantum mechanics provides a simple interpretation of the experimental findings. A silver atom has a spin of 1/2 unit of angular momentum $h/2\pi$. The atoms in one beam have a spin orientation quantum number (for the z-direction) of $+1/2$ and in the other of $-1/2$. Individual atoms can be detected, and it is certain that whether an atom goes into the $+1/2$ beam or the $-1/2$ beam is determined by just the same kind of random process which Schrödinger invoked for his radioactive decay. No experiment can be devised to tell beforehand what an individual atom will do. I think it is highly desirable to reformulate the Cat paradox in terms of the two level problem. None of the physics is lost, and unclear thinking is not encouraged.

Furthermore, as a bonus, double or compound Stern–Gerlach

experiments can be made and analyzed. The atoms of the upper beam might be passed into another inhomogeneous magnetic field which will split that beam into two beams. The numbers of particles in the new beams depend on the angles of the various fields in just the way uniquely characteristic of and described by quantum mechanics.

In 1949, shortly before he died, Einstein[31] explained quite clearly why he did not like probability in quantum mechanics. He wanted a theory of the radioactive nuclei which could predict ahead of time when a nucleus would decay. Perhaps he had the hope that a 'hidden variable' theory could be found. He would have at least as hard a problem with the silver atoms of the Stern–Gerlach experiment. Now, through the theoretical work of John Bell[32] and the experiments of Alain Aspect,[33] we know that what Einstein wanted cannot be had. Quantum mechanics really does work in simple cases, and probability must be used in the interpretation. Our genetic heritage and social pressures make us feel that things are deterministic. However, on an atomic level, they simply are not. In principle, we could apply quantum descriptions to large scale phenomena, but, in general, we can't because we are unwilling or unable to adhere to the rules of the game. If such a system is observed, it is disturbed. One can try to take the effect of the disturbance into account, but this inevitably introduces an element of chance. I think it is high time that we realize that this is inevitable and learn to enjoy it.

References

1 C. Lanczos, On a field representation of the new quantum mechanics, *Zeit. f. Physik* **35**, 812–30 (1926). See also, B. L. van der Waerden in *The Physicist's Conception of Nature*, pp. 276–93, ed. J. Mehra, Reidel, Dordrecht (1973).

2 E. H. Kennard, Quantum mechanics of simple types of motion, *Zeit. f. Physik* **44**, 326–52 (1927). Kennard's principle of 'transitivity' can be used to break up the time interval (t_1, t_2) into a large number of small time intervals. A simple consideration converts the Hamiltonian in Kennard's exponential function into the Lagrangian in Feynman's path integral.

3 R. P. Feynman & A. R. Hibbs, *Quantum Mechanics and Path Integrals*, McGraw-Hill, New York (1965).

4 L. D. Landau, The damping problem in wave mechanics, *Zeit. f. Physik* **45**, 430–41 (1927).

Willis E. Lamb, Jr.

5 J. von Neumann, Thermodynamics of quantum mechanical ensembles, *Goettingen Nachrichten* **27**, 273–91 (1927).

6 P. A. M. Dirac, Basis of statistical quantum mechanics, *Proc. Camb. Phil. Soc.* **25**, 62–6 (1929).

7 W. Heisenberg, On the intuitive content of the quantum mechanical kinematics and mechanics, *Zeit. f. Physik* **43**, 172–98 (1927).

8 H. P. Robertson, The uncertainty principle, *Phys. Rev.* **34**, 163–4 (1929).

9 E. Schrödinger, *On Heisenberg's Uncertainty Principle*, pp. 296–303, Sitzungsber. Preuss. Akad. Wiss. (1930).

10 W. Heisenberg, *The Physical Principles of the Quantum Theory*, Chicago Univ. Press (1930).

11 E. Schrödinger, *Collected Papers on Wave Mechanics*, Blackie, Glasgow (1928).

12 E. Schrödinger, On the dynamics of elastically coupled point systems, *Ann. d. Physik* **44**, 916–34 (1944).

13 P. A. M. Dirac, The quantum theory of the emission and absorption of radiation, *Proc. Roy. Soc.* **A114**, 243–65 (1928).

14 W. Heisenberg & W. Pauli, On quantum dynamics of wave fields, *Zeit. f. Physik* **56**, 1–61 (1929).

15 P. A. M. Dirac, The quantum theory of the electron, *Proc. Roy. Soc.* **117**, 610–24 (1928).

16 J. Gribben, *In Search of Schrödinger's Cat*, Bantam (1984).

17 R. Peierls, Prowling around quantum mechanics: in search of Schrödinger's cat, Book Review, *New Scientist*, p. 45 (December 13, 1984).

18 R. MacCormmach, Schrödinger's mysterious cat: It may be alive, it may be dead, it may be both, Book Review, *Science Digest*, p. 78 (January 1985).

19 H. Everett III, Relative state formulation of quantum mechanics, *Rev. Mod. Phys.* **29**, 454–62 (1957).

20 A. Einstein, B. Podolsky & N. Rosen, Can quantum-mechanical description of physical reality be considered complete?, *Phys. Rev.* **47**, 777–80 (1935).

21 E. Schrödinger, The present situation in quantum mechanics, *Naturwissenschaften* **23**, 807–12, 823–8, 844–9 (1935).

22 E. Schrödinger, *What is Life?*, Cambridge Univ. Press (1944); also reprinted, Doubleday (1956).

23 P. A. M. Dirac, *The Principles of Quantum Mechanics*, 1st edn., Oxford Univ. Press (1930).

24 J. von Neumann, *Mathematical Foundations of the Quantum Mechanics*, Springer-Verlag (1933); also reprinted in translation, Princeton (1955).

25 W. E. Lamb Jr., An operational interpretation of non-relativistic quantum mechanics, *Physics Today*, pp. 23–8 (22 April, 1969).

26 W. E. Lamb Jr., Von Neumann's reduction of the wave function, in

Proceedings of the Fourth International Conference on the Unity of the Sciences, pp. 297–303, International Cultural Foundation (1975).

27 W. E. Lamb Jr., Remarks on the interpretation of quantum mechanics, in *Ta-Yu Wu Festschrift: Science of Matter*, pp. 1–8, ed. S. Fujita, Gordon and Breach (1979).

28 W. E. Lamb Jr., A letter to Kip Thorne, in *CCNY Physics Symposium in Celebration of Melvin Lax's Sixtieth Birthday*, pp. 40–3, ed. H. Falk, CCNY, New York (1983).

29 W. Pauli, The general principles of wave mechanics, *Handbuch der Physik*, Vol. 24/1, pp. 83–272, eds. H. Geiger & K. Scheel, Springer-Verlag (1933).

30 See ref. 26.

31 A. Einstein, Remarks concerning the essays brought together in this co-operative volume, in *Albert Einstein: Philosopher-Scientist*, 3rd edn., pp. 655–88, ed. P. A. Schilpp, Open Court, La Salle, IL (1969).

32 J. S. Bell, On the problem of hidden variables in quantum mechanics, *Rev. Mod. Phys.* **38**, 447–52 (1966).

33 A. Aspect, P. Grangier & G. Roger, Experimental tests of realistic local theories via Bell's theorem, *Phys. Rev. Lett.* **47**, 460–3 (1981).

22

Dirac and finite field theories

Abdus Salam
International Centre for Theoretical Physics, Trieste

1

Paul Adrien Maurice Dirac was undoubtedly one of the greatest physicists of this or any century. In three decisive years – 1925, 1926 and 1927 – with three papers, he laid the foundations, first, of quantum physics, second, of the quantum theory of fields, and, third, of the theory of elementary particles, with his famous equation of the electron. No man except Einstein has had such a decisive influence, in so short a time, on the course of physics in this century. For me, personally, Dirac represented the highest reaches of personal integrity of anyone I have ever known. Thus, on this Commemorative Day, with sadness for his passing, and with deeply felt sympathy for his bereaved family, let me begin with sentiments of rejoicing for his life and for our good fortune in having known him.

In speaking of Dirac's work today, I wish to take as my subject the area of renormalization. Even though Dirac and Kramers were the first to emphasize the physical necessity of the concept of (finite) renormalization (of the electron's mass), Dirac never approved of our use of this beautiful idea of renormalization to hide away the infinities which appear in perturbation calculations in quantum electrodynamics. He believed in and always advocated that, eventually, a finite field theory would be discovered for all processes.

262

Dirac and finite field theories

My generation of theoretical physicists was brought up on the work of Tomonaga, Schwinger, Feynman, and, in particular, of Dyson. Dyson proved that all infinities in quantum electrodynamics in each perturbative order could be absorbed into a renormalization of electron mass and charge. This was an important result. Thus, at the price of not being able to compute these two quantities* alone, all scattering processes in quantum electrodynamics could be made finite. This, in Dyson's view, was a small price to pay for a resolution of the field theoretic infinity-problem. My generation avidly bought this idea, but not so Dirac.

Whilst recognising that such absorption of infinities through a renormalization of mass and charge could be a temporary expedient, Dirac always insisted that there is no place for infinities in a fundamental field theory. He felt strongly that one should keep searching for a basic amelioration of the infinity problem.

It now appears that there is, indeed, a class of field theories which are perturbatively finite to all orders. If renormalizations of coupling constants and masses are physically necessary, these would only be finite renormalizations. The hitherto discovered field theories of this type are non-Abelian gauge theories of the extended supersymmetric type. Whether such theories are physically relevant is not yet known, but they are mathematically elegant and, without doubt, satisfy the criterion of beauty which Dirac always advocated.

Even more important, there has recently been developed a local field theory of extended one-dimensional objects (strings).† There is the promise – brought to a near proof – that closed-string supersymmetric field theories, whose long-range excitations must contain quantum gravity (as well as Yang–Mills excitations describing electro-nuclear interactions of matter) may give rise to finite matrix elements. If this conjecture is finally proved, and if these theories prove to be physically relevant, Dirac would be fully vindicated.

---

* There is a third infinity associated with the wave function of the electron, but this is gauge dependent and one can find gauges in which this quantity could be rendered finite.
† Dirac, one may recall, was one of the pioneers of field theories of (two-dimensional) extended objects.

2

The finite supersymmetric field theories at present known fall into two classes:

(1) $N = 2$ Yang–Mills supersymmetric theories, where the Yang–Mills gauge particles (plus four-component gauginos and complex scalars) belong to the adjoint representation of the gauge group, e.g., SU(n). This Yang–Nills $N = 2$ supersymmetric multiplet interacts with $2n$ fundamental representation of $N = 2$ matter multiplets. There is no coupling of matter multiplets among themselves.

(2) The second type of finite theories are the $N = 4$ Yang–Mills supersymmetric field theories.

Let me describe the $N = 4$ Yang–Mills theories first. These are theories with an internal symmetry group SU(n). (The symmetry group could be any non-Abelian group G, but SU(n) is a good example.) This theory contains $(n^2 - 1)$ Yang–Mills spin-one objects in the adjoint representation of SU(n), $4(n^2 - 1)$ two-component chiral fermions ψ_α and $3(n^2 - 1)$ complex scalar particles $H_{[\alpha\beta]}$, where $H_{41} = H^{32}$, etc. with the index $\alpha = 1, 2, 3, 4$ characterizing the underlying $N = 4$ supersymmetry.

The interaction Lagrangian is given by

$$\int d^4x \left[-\tfrac{1}{4}F_{\mu\nu}^2 - \bar{\psi}^\alpha i \not\nabla \psi_\alpha + \tfrac{1}{4}\nabla^\mu \bar{H}^{\alpha\beta} \cdot \nabla_\mu H_{\alpha\beta} \right.$$

$$+ \frac{g}{\sqrt{2}}(\bar{H}^{\alpha\beta} \cdot \psi_\alpha^T \times \mathscr{C}^{-1}\psi_\beta + \text{h.c.})$$

$$\left. + \frac{1}{2}\sum_\alpha M_\alpha(a)(\psi_\alpha^T \cdot \mathscr{C}^{-1}\psi_\alpha + \text{h.c.}) - V(H) \right], \tag{1}$$

where $V(H)$ is

$$V(H) = M^2(H_{14})|H_{14}|^2 + M^2(H_{24})|H_{24}|^2 + M^2(H_{34})|H_{34}|^2$$

$$+ \frac{\sqrt{2}}{6}g\sum_\alpha M_\alpha(a)\varepsilon^{\alpha\beta\gamma\delta}H_{\beta\alpha} \cdot H_{\gamma\alpha} \times H_{\delta\alpha} + \text{h.c.}$$

$$+ \frac{g^2}{16}\bar{H}^{\alpha\beta} \times H^{\gamma\delta} \cdot H_{\alpha\beta} \times H_{\gamma\delta}. \tag{2}$$

Dirac and finite field theories

The Lagrangian contains seven masses (four for the fermions ψ_α and three for the complex scalars H_{14}, H_{24}, H_{34}). The inclusion of these masses breaks supersymmetry softly. It can be shown that the Green's functions of the theory exhibit no ultraviolet infinities, provided the following conditions are satisfied.

(1) The masses of the particles concerned obey the relation:

$$M^2(H_{14}) + M^2(H_{24}) + M^2(H_{34})$$
$$= M^2(\psi_1) + M^2(\psi_2) + M^2(\psi_3) + M^2(\psi_4).$$

(2) The Green's functions must be computed in the light cone gauge. If a general gauge is employed, there will, of course, be no ultraviolet infinities on-shell, but such infinities may arise for the off-shell wave function renormalizations.

(3) If we restrict n to $n = 2$ (i.e., the internal gauge symmetry group is SU(2)), there are no infrared infinities in the theory either, i.e., the theory is completely finite in the light cone gauge, both for ultraviolet and infrared infinities.

(4) For $n > 2$, some infrared infinities may remain associated with the SU($n - 2$) unbroken Yang–Mills symmetries due to the masslessness of the Yang–Mills vector particles, associated with the SU($n - 2$).

This result (M. A. Namazie, J. Strathdee & Abdus Salam, *Phys. Rev.* **D28**, 1481 (1983)) is the most powerful to date of a series of results which started with the advent of supersymmetry in the latter part of 1973. I shall trace some of the steps in this development.

(1) For supersymmetric theories (with the same number of fermion and boson modes), Weiss and Zumino showed that there was a mutual cancellation of loop infinities between fermions and bosons, such that the vertex-part infinities cancelled in a Yukawa type of supersymmetry theory. The only uncancelled infinities in such theories were those associated with the mass and the wave function renormalizations!

(2) For Yang–Mills supersymmetric theories, however, the wave function renormalizations are infinite, and so are the vertex renormalizations. This implies a charge renormalization. In an axial gauge, this renormalization is associated with the wave function renormalization of the gauge particles, i.e., $e_R = Z_3^{1/2} e$.

(3) For $N = 4$ Yang–Mills theory without mass, it was noted by L. V. Avdeev, O. V. Tarasov & A. A. Vladimirov (*Phys. Lett.* **96B**, 94 (1980));

M. T. Grisaru, M. Roček & W. Siegel (*Phys. Rev. Lett.* **45**, 1063 (1980)); and W. Caswell & D. Zanon (*Phys. Lett.* **100B**, 152 (1980)) that, up to three loops, the β-function (which is a gauge invariant quantity and which determines whether there is an infinity associated with charge renormalization), is actually zero (i.e., $Z_3 = 1$ in $e_R = Z_3^{1/2} e$).

(4) Suggestive arguments were given to all orders for the vanishing of the β-function by S. Ferrara & B. Zumino (unpublished), M. Sohnius & P. West (*Phys. Lett.* **100B**, 245 (1981)), and K. Stelle in *Quantum Structure of Space and Time*, p. 337, eds. M. J. Duff & C. J. Isham, Cambridge Univ. Press (1982).

(5) These arguments were made quantitative and extended to include Green's functions to all orders for the massless theory by S. Mandelstam (*Nucl. Phys.* **B213**, 149 (1983) and *21st Int. Conf. on High Energy Physics, Paris* (1982)) and L. Brink *et al.* (*Phys. Lett.* **123B**, 323 (1983)) in the light cone gauge. Not only was the gauge invariant charge renormalization finite, but in this gauge, all the Zs were also finite.

(6) There was, however, a worry that the massless, fully supersymmetric theory may, after all, be a free field theory. One needed to see if the results on finiteness would survive the breaking of supersymmetry implied by the addition of mass terms. Such additions of mass terms are expected to be gentle, so far as the infinities are concerned. This led to the investigation by Namazie, Strathdee, and Salam mentioned above. We discovered that, if the supertrace of M^2 is equal to zero, and a specific set of cubic terms of the type indicated in (1) and (2) are added to the Lagrangian (together with the mass terms), the theory retains its finiteness (in the light cone gauge) to all orders.

(7) The infrared finiteness for the case of $n = 2$ (i.e., internal symmetry group SU(2)) comes about because it turns out that the theory possesses a vacuum solution where the internal symmetry group (SU(2)) is spontaneously broken such that all three vector mesons in the theory acquire spontaneously generated masses. Thus, no infrared infinities can arise. Namazie has gone further and shown that if $n = 3$ (i.e., the internal symmetry group is SU(3)), the symmetry breaks down to U(1). His conjecture is that SU(n) may be expected to break down to SU($n-2$) × U(1) so that $(n-2)^2$ vector mesons remain massless, leading to the standard infrared problems (familiar in QCD). Results similar to ours were obtained by J. G. Taylor and A. Parkes and P. West. (These authors did not work in the light cone gauge.)

3

In addition to the $N = 4$ theory described so far, there is the set of $N = 2$ supersymmetric Yang–Mills theories which, for charge renormalization, are also finite. This comes about because, as M. T. Grisaru & W. Siegel showed in 1982 (*Nucl. Phys.* **B201**, 292 (1982)), for β-functions, the only surviving infinities are expected to exist for one loop graphs. All higher loops are finite – provided there exists an extended superfield formalism for such a theory. Such a formalism was, indeed, recently provided by P. Howe, K. Stelle & P. Townsend (*Phys. Lett.* **124B**, 55 (1983)). P. Howe, K. Stelle and P. West suggested that, if a class of matter multiplets could be found such that their infinities and the infinities from the gauge sector cancelled in the *one loop* approximation, the β-function would be zero to all orders. There would, therefore, be no charge renormalization.

Precisely, this happens for SU(n) internal symmetry, provided one takes $2n$ ($N = 2$) matter multiplets in the fundamental representation (or alternatively, one $N = 2$ matter multiplet in the adjoint representation). In an axial or a background gauge, the wave-function renormalizations are also finite. (See Appendix 1.)

As I said earlier, we do not know if these theories are physically relevant. J. Pati & Abdus Salam (*Nucl. Phys.* **B214**, 109 (1984), section V) have suggested that such theories may describe preons, the basic constituents of quarks and leptons. Time and experimentation will tell if this conjecture is justified. In the meanwhile, we must salute the physical insight of Dirac's insistence that one must continue looking for finite field theories, rather than be reconciled with ultraviolet infinities – even though these may refer to only a few physical quantities like mass and charge.

Finally, $N = 3$ supersymmetric Yang–Mills has also been shown finite (E. Ahmed & J. G. Taylor, King's College Preprint) using the harmonic superspace techniques of V. Ogievetsky, E. Sokatchev, and their collaborators.

4

I have not discussed the supergravity theories so far. When these were invented in 1976, the hope was that they might make infinities of the gravity theory tractable. Such a hope was entertained particularly for extended supergravity theory $N = 8$ $(d = 4)$. Unhappily, this hope has not been justified yet, and perhaps never will be. In Appendix 2 is given the present situation as reviewed by Martin Sohnius.

A new possibility for renormalization of gravity, however, has opened up from string theories and, in particular, for closed superstring theories in 10 dimensions based on the group $E_8 \times E_8$. Mandelstam, in particular, has recently given powerful arguments to show that all closed string theories (which include Einstein's gravity) are likely to be finite to all orders. This result will not apply to open string theories which are expected to be renormalizable only.

The string theories are based on 2-dimensional Nambu–Gotto theories, for which Alvarez-Gaumé and D. Z. Freedman had shown (following the earlier work of Freedman and Townsend) that the $N = 4$ supersymmetric non-linear sigma models are finite. This result has been recently extended by Hull to $N = 1$ (and $N = 2$) supersymmetric non-linear sigma models that are defined on manifolds which are Ricci flat (D. Z. Freedman & P. Townsend, *Nucl. Phys.* **B177**, 282 (1981); Alvarez-Gaumé & D. Z. Freedman, *Comm. Math. Phys.* **80**, 443 (1981) and *Phys. Rev.* **D15**, 846 (1980); Alvarez-Gaumé, D. Z. Freedman & S. Mukhi, *Ann. Phys.* **134**, 85 (1985)). In the construction of string theories, these basic finite 2-dimensional string theories are destined to play an important role.

5

I would like to conclude with a number of personal recollections about Dirac.

I once asked Dirac what he thought his greatest contribution to physics was. He said: 'The Poisson bracket.' I was surprised, because I thought he would speak of the electron equation. When I asked him to elaborate on this, he said that, after a long search for the quantum analogue of the Poisson bracket, it struck him that, if there are two non-commutative operators – A and B, then the quantity $AB - BA$ would have all the properties of a classical Poisson bracket. However, on the Sunday on which he had made this discovery, he had no texts on dynamics available, and it was an anxious wait until Monday morning when he could finally go to the Library and check whether his expression, indeed, satisfied all conditions a quantum Poisson should.

But, with characteristic modesty, he added after a pause that, for a long time, he felt ecstatic and pleased, till he found essentially the same remark made by Hamilton as a footnote in one of his papers written in the last century.

I once unwisely criticized Eddington in Dirac's presence. My remarks were the result of exasperation with Eddington's 'Fundamental Theory'. I believe I said that, if Eddington were not a professor at Cambridge, he would not have had his book published. Dirac made the remark (which I have appreciated deeply later): 'One must not judge a man's worth from his poorer work; one must always judge him by the best he has done.'

Once, Dirac asked me whether I thought geometrically or algebraically? I said I did not know what he meant, could he tell me how he, himself, thought. He said his thinking was geometrical. I was taken aback by this because Dirac, with his transformation theory, represented for my generation the algebraic movement in physics *par excellence*. So, I said: 'I still don't understand.' He said: 'I will ask you a question. How do you picture the de Sitter space?' I said, 'I write down the metric and then think about the structure of the terms in the expression.' He said, 'Precisely as I thought. You think algebraically, as most people from the Indian sub-continent do. I picture, without effort, the de Sitter space as a four-dimensional surface in a five-dimensional space.'

Most of these conversations took place at High Table at St. John's College when, every Tuesday, he drove into College in his two-seater for dinner. It was my very great pleasure to sit next to him for several terms at that High Table, and I always anticipated the pleasure of those evenings.

Dirac was always very economical with words. Perhaps, one of the best recorded stories is of a newspaper interview that Dirac gave to a Wisconsin journal and which is recounted in the paper by Laurie M. Brown and Helmut Rechenberg (Chap. 12).

The residents of Madison, other than the physicists at the University, learned about Dirac from a local newspaper interview:

I been hearing about a fellow they have up at the U. this spring – a mathematical physicist, or something, they call him – who is pushing Sir Isaac Newton,

Abdus Salam

Einstein and all the others off the front page ... His name is Dirac and he is an Englishman ... So the other afternoon I knocks at the door of Dr. Dirac's office in Sterling Hall and a pleasant voice says 'Come in.' And I want to say here and now that this sentence 'come in' was about the longest one emitted by the doctor during our interview.

I found the Doctor a tall youngish-looking man, and the minute I seen the twinkle in his eye I knew I was going to like him ... he did not seem to be at all busy. Why if I went to interview an American scientist of his class ... he would blow in carrying a big briefcase, and while he talked he would be pulling lecture notes, proof, reprints, books, manuscripts, or what have you, out of his bag. Dirac is different. He seems to have all the time there is in the world and his heaviest work is looking out the window ...

'Professor,' says I, 'I notice you have quite a few letters in front of your last name. Do they stand for anything in particular?'

'No,' says he.

'... Fine,' says I ... 'Now Doctor will you give me in a few words the low-down on all your investigations?'

'No,' says he.

I went on: 'Do you go to the movies?'

'Yes,' says he.

'When?' says I.

'In 1920 – perhaps also 1930,' says he.

'... And now I want to ask you something more: They tell me that you and Einstein are the only ones who can really understand each other ... Do you ever run across a fellow that even you can't understand?'

'Yes,' says he.

'... Do you mind releasing to me who he is?'

'Weyl,' says he.

When at Trieste, we had the idea of celebrating his 70th birthday, he was most reluctant. In the end, Mrs. Dirac convinced him otherwise and he agreed. One of the best evenings in my life was the Banquet evening for that celebration when, in his presence, those present told their recollections of Dirac.*

In Trieste, we constructed 'Scala Dirac' – the Dirac Stairs which take visitors from the Centre to the Park and are built along the face of a steep slope. These stairs were named Scala Dirac, because my secretary once saw Dirac trying to scale down that slope. He could not quite negotiate

* Published in J. Mehra's volume, *The Physicist's Conception of Nature*, Reidel, Dordrecht and Boston (1973).

it because it was so steep. Dirac then happily started to slide down. We had to build the stairs and to name them, then, after him.

The Centre has now instituted a Dirac Prize to be given in his honour on his birthday – 8 August – every year, for the highest achievements in theoretical physics.

I will conclude with a story of Dirac and Feynman that, perhaps, will convey to you, in Feynman's words, what we all thought of Dirac. I was a witness to it at the 1961 Solvay Conference. Those of you who have attended the old Solvay Conferences will know that, at least then, one sat at long tables that were arranged as if one was sitting to pray. Like a Quaker gathering, there was no fixed Agenda; the expectation – seldom belied – was that someone would be moved to start off the discussion spontaneously.

At the 1961 Conference, I was sitting at one of these long tables next to Dirac, waiting for the session to start, when Feynman came and sat down opposite. Feynman extended his hand towards Dirac and said: 'I am Feynman.' It was clear from his tone that it was the first time they had met. Dirac extended his hand and said: 'I am Dirac.' There was silence, which from Feynman was rather remarkable. Then Feynman, like a schoolboy in the presence of a Master, said to Dirac: 'It must have felt good to have invented that equation.' And Dirac said: 'But that was a long time ago.' Silence again. To break this, Dirac asked Feynman, 'What are you yourself working on?' Feynman said: 'Meson theories' and Dirac said: 'Are you trying to invent a similar equation?' Feynman said: 'That would be very difficult.' And Dirac, in an anxious voice, said: 'But one must try.' At that point, the conversation finished because the meeting had started.

Appendix 1

Finiteness of $N = 2$ theories

The condition for one loop finiteness which guarantees finiteness to all orders is

$$T(R) = C_2(G),$$

where $T(R)$ and $C_2(G)$ are the Dynkin index of matter multiplets in representation R and the quadratic Casimir operator for the adjoint representation of gauge group G. A complete list of finite field theories can be

Abdus Salam

readily obtained. (See, for example, I. G. Koh & S. Rajpoot, *Phys. Lett.* **135B**, 397 (1984).) If matter multiplets are assigned solely to the adjoint representations, the $N = 2$ theory coincides with $N = 4$ theory.

For SU(n) internal symmetry, $N = 2$ supersymmetric Yang–Mills theories with the following representations for matter multiplets are finite:

(a) $2n\ \square$ for all SU(n),

(b) $(n-2)\square + \square\square$ or $[2n - p(n-2)]\square + p\ \begin{array}{c}\square\\\square\end{array}$ for SU(n) with $n \geqslant 3$,

(c) $\begin{array}{c}\square\\\square\end{array} + \square\square$ for SU(n) with $n \geqslant 4$,

(d) $2\ \begin{array}{c}\square\\\square\\\square\end{array}$ or $2\ \square + \begin{array}{c}\square\\\square\end{array} + \begin{array}{c}\square\\\square\\\square\end{array}$ for SU(6), and

(e) $1/2(9n - n^2 - 6)\ \square + \begin{array}{c}\square\\\square\end{array}$ for SU(n) with $6 \leqslant n \leqslant 8$.

Here, the coefficients in front of Young tableaux denote the multiplicities and are positive.

For SO(n) group, the representations for the matter multiplets of finite field theories are the following:

(a) $(n - 2)$ repetitions of **n** (fundamental) representation.

(b) If the matter multiplets are assigned to spinor representations of SO(n), the following four cases, 6(**4**) of SO(5), 5(**8**) of SO(7), 6(**8**) of SO(8) and 4(**16**) of SO(10) satisfy the condition for finiteness, with multiplicities shown in front of dimensionality.

(c) The representations with n_f repetitions of the fundamental representation of dimension n and with n_s repetitions of the spinorial representation of dimension $2^{[n/2]}$ for odd n, and $s^{[n/2]-1}$ for even n satisfy the finiteness condition with the following (n_f, n_s) combinations:

SO(5) $\quad (n_f, n_s) = (2, 2), (1, 4)$

SO(7) $\quad (n_f, n_s) = (4, 1), (3, 2), (2, 3), (1, 4)$

SO(9) $\quad (n_f, n_s) = (5, 1), (3, 2), (1, 3)$

SO(10) $\quad (n_f, n_s) = (6, 1), (4, 2), (2, 3)$

SO(12) $\quad (n_f, n_s) = (6, 1), (2, 2)$

SO(14) $\quad (n_f, n_s) = (4, 1).$

272

The representations of matter multiplets allowed for Sp(2m) group are:
(a) $(2m + 2)$ repetitions of **2m** fundamental representation,
(b) four repetitions of **2m** and one second-rank antisymmetric representation of dimension $m(2m - 1)$.
To give more details of $N = 2$ theories in exceptional groups, the following representations of matter multiplets form $N = 2$ finite field theories:

(a) four times **27** in E_6
(b) three times **56** in E_7
(c) three times **26** in F_4
(d) four times **7** in G_2.

It is interesting to note that the only representation satisfying the finiteness condition for E_8 is the adjoint representation. Thus, finiteness for E_8 gauge group requires the $N = 4$ theory.

Appendix 2

Quantum gravity

$$K = \sqrt{8\pi G} \simeq 5 \times 10^{-19} \, \text{GeV}^{-1},$$

$$K = \frac{1}{\text{mass}} \Rightarrow \text{non-renormalizable}.$$

Contrast: Yang–Mills:
  Renormalizable, only a few graphs diverge.
  Supersymmetry can make these finite.
Gravity:
  Non-renormalizable, infinitely many divergences.
  Infinitely many cancellations required.
  Can (local) SUSY achieve this?

Original hope: Non-existence of counterterms
  Example: '*pure gravity*' (no matter)
  1-loop: General form of counter term

$$\alpha R_{\mu\nu\rho\sigma} R^{\mu\nu\rho\sigma} + \beta R_{\mu\nu} R^{\mu\nu} + \gamma R^2$$

$$\Rightarrow \text{No surviving counterterm, 1-loop finite}$$

if one uses Gauss–Bonnet theorem and the equation of motion $R_{\mu\nu} = 0$.
For 2-loops: $\alpha R_{\mu\nu\rho\sigma} R^{\rho\sigma\kappa\lambda} R^{\mu\nu}_{\kappa\lambda}$ remains $\neq 0$ on-shell.
  This term does *not* have a supersymmetric extension.

273

One can show that *Supergravity* is 2-loop finite if 1-loop finite.
This is true if no matter is present;
but:
For 3-loops: $\alpha((\text{Bel–Robinson})^2 + \text{SUSY-extension})$ exists for $N \leqslant 4$.
Also 7-loops: $\alpha((\text{Weyl})^8 + \text{SUSY-extension})$ exists for $N = 8$.
It has been conjectured that 3-loop term probable for $N = 8$.

Remaining hope: $\alpha = 0$
This requires more miracles than SUSY is likely to deliver (no α has yet been
calculated at 3-loop level).

Appendix 3

$N = 4$ Super – Yang–Mills with soft breaking terms

Fields:

A_μ		$n^2 - 1$,
ψ_α	$\alpha, \beta = 1, \ldots, 4$ (internal index)	$4(n^2 - 1)$,
$H_{[\alpha\beta]}$	H complex, $H_{41} = \bar{H}^{32}$, etc.	$6(n^2 - 1)$,
	for SU(n) gauge group.	

Lagrangian:

$$\mathscr{L} = -\tfrac{1}{4}F_{\mu\nu}^2 - \bar{\psi}^\alpha i\nabla\!\!\!\!/\,\psi_\alpha + \tfrac{1}{4}\nabla^\mu \bar{H}^{\alpha\beta} \cdot \nabla_\mu H_{\alpha\beta}$$

$$+ -\frac{g}{\sqrt{2}}(\bar{H}^{\alpha\beta} \cdot \psi_\alpha^T \times \mathscr{C}^{-1}\psi_\beta + \text{h.c.})$$

$$+ -\frac{1}{2}\sum_\alpha M_\alpha(a)(\psi_\alpha^T \cdot \mathscr{C}^{-1}\psi_\alpha + \text{h.c.}) - V(H), \qquad (1)$$

$$V(H) = M^2(H_{14})|H_{14}|^2 + M^2(H_{24})|H_{24}|^2 + M^2(H_{34})|H_{34}|^2$$

$$+ -\frac{\sqrt{2}}{6}g\sum_\alpha [M_\alpha(\alpha)\varepsilon^{\alpha\beta\gamma\delta}H_{\beta\alpha} \cdot H_{\gamma\alpha} \times H_{\delta\alpha} + \text{h.c.}]$$

$$+ \frac{g^2}{16}\bar{H}^{\alpha\beta} \times H^{\gamma\delta} \cdot H_{\alpha\beta} \times H_{\gamma\delta}. \qquad (2)$$

Seven masses: four for fermions, three for Hs.
(Scalar)^3 – coupling which depends on fermion masses.

Finite if:

(1) $M^2(H_{14}) + M^2(H_{24}) + M^2(H_{34})$

$$= M^2(\psi_1) + M^2(\psi_2) + M^2(\psi_3) + M^2(\psi_4).$$

(2) light-cone gauge

(otherwise: infinite off-shell wave function renorm.).

(3) No IR-infinities if $n = 2$ (i.e., for SU(2) – Yang–Mills)

(otherwise: IR-infinities ass. with unbroken sector of gauge group).

(From Namazie, Strathdee & Salam, *Phys. Rev.* **D28**, 1481 (1983).)

23

Dirac's influence on unified field theory

Behram N. Kursunoglu
University of Miami

1 Introduction

The last paragraph of the following letter from Dirac concerns his attempt to unify gravitation and electromagnetism based on Weyl's generalization of Einstein's relativistic theory of gravitation. Dirac's approach was contained in a paper, entitled 'The Geometrical Nature of Space and Time,' which he presented at the 1974 Coral Gables Orbis Scientiae.[1] Dirac states that, 'The passage from Euclidean to Riemann space was so successful. Should one not try to take another step in the same direction? Passing to Riemann space provided one with an excellent description of the gravitational field. Now there is another field, the electromagnetic field. Both fields involve long-range forces, which fall off according to the law r^{-2}. This distinguishes them from the other fields of physics, which are specially important for atomic physics, involving short-range forces, which fall off according to the law $\exp(-\kappa r)$. Should one not treat the electromagnetic field like the gravitational, and try to explain it in terms of a more general geometry? This leads to the problem of the unification of the gravitational and electromagnetic fields.' For further details of Dirac's approach, I would like to refer the reader to his paper.[1]

I attended Dirac's quantum mechanics course at Cambridge University as a research student from 1949–52 and, thus, belonged to a

276

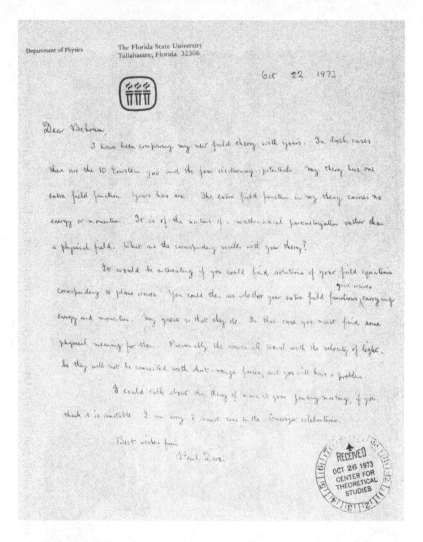

privileged group of students at that time. I was not officially assigned to Dirac's supervision as a research student, but I visited him more often than did most of his research students. He always took great interest in Einstein's unified field theories. I was especially pleased to show him Einstein's and Schrödinger's letters replying to my letters to them on the nonsymmetric generalization of the general theory of relativity. His comments on these letters and on my own work on unified field theory have been a source of inspiration for me. Our relationship, which was

P. A. M. Dirac asking a question at the 1971 Orbis Scientiae at the Center for Theoretical Studies, University of Miami.

interrupted after I left Cambridge at the end of 1952, was resumed when I invited the Diracs to visit the Center for Theoretical Studies in 1968. He accepted a one-week-only stay in Florida but, happily, stayed in Coral Gables until 1972, when he accepted an offer from Florida State University in Tallahassee and stayed there until his death in October 1984. He visited the Center every year for two weeks and was the first speaker in our annual Orbis Scientiae, the last of which in 1983 was dedicated to his 80th year.

I resumed my work on unified field theory in 1969 when Dirac arrived in Coral Gables and have further developed my own version of the nonsymmetric generalization of general relativity, alternative to those of Einstein and Schrödinger. All three versions of the nonsymmetric theory were discussed by Einstein's assistant, Brira Kaufman,[2] and later included in a book on Einstein by A. Pais.[3] The first two paragraphs of Dirac's preceding letter refer to this theory of mine and have led me to answer his question and concern that I may have a problem regarding the possibility that even short-range forces might propagate with the velocity of light. The following is a shortened presentation concerning the propagation of short-range forces, which travel with a velocity less than that of light. In fact, I traveled to Tallahassee in December of 1973 to give a talk on this subject and fully convinced Dirac of this important result. It gave me great joy and satisfaction to gain his approval on this.

2 Propagation of short-range forces

In this section, we shall demonstrate that the propagation of short-range field takes place with a velocity less than that of light. The propagation of waves in a nonlinear field theory can be studied in an indirect way. In fact, the velocity of energy and momentum-carrying waves can provide a good test for the physical validity of the theory. Here, we shall be concerned only with nongravitational waves.

In order to study the nature of the short-range forces in this theory, we shall consider, of the sixteen-field equations, only the linearized form of the four field equations[4]

$$\mathbf{R}_{\mu\nu,\rho} + \mathbf{R}_{\nu\rho,\mu} + \mathbf{R}_{\rho\mu,\nu} + r_0^{-2} I_{\mu\nu\rho} = 0, \qquad (2.1)$$

for the nonsymmetric field variables $\mathbf{g}_{\mu\nu} = g_{\mu\nu} + q^{-1}\Phi_{\mu\nu}$, where the magnetic current density

$$I_{\mu\nu\rho} = \Phi_{\mu\nu,\rho} + \Phi_{\nu\rho,\mu} + \Phi_{\rho\mu,\nu},$$

is related to its dual vector ∂^{μ} according to

$$I_{\mu\nu\rho} = \varepsilon_{\mu\nu\rho\sigma}\partial^{\sigma}.$$

The integrated form of (2.1),

$$r_0^2 \mathbf{R}_{\underset{\vee}{\mu\nu}} + \Phi_{\mu\nu} = F_{\mu\nu} = \partial_{\mu}A_{\nu} - \partial_{\nu}A_{\mu},$$

contains, for short distances, a modification of electrodynamics. We shall study the general behavior of (2.1) for weak fields at distances large, compared to r_0, where $\mathbf{R}_{\underset{\vee}{\mu\nu}}$ represents the antisymmetric part of the nonsymmetric curvature tensor $\mathbf{R}_{\mu\nu}$ and where the constant length r_0 and the constant q (with the dimensions of an electric field) are related according to

$$r_0^2 q^2 = \frac{c^4}{2G}. \tag{2.2}$$

Thus, in an approximately Galilean region, i.e., at distances

$$r \gg r_0, \tag{2.3}$$

we can neglect (i) the gravitational field and (ii) all the terms containing cubic or higher powers at the constant q^{-1}. Under these assumptions, the above-mentioned field equations can be linearized into a wave equation

$$\left(\nabla^2 - \frac{1}{c^2}\frac{\partial^2}{\partial t^2} + \kappa^2\right)\partial^{\mu} = 0, \tag{2.4}$$

for the particle's magnetic current density ∂^{μ}. In (2.4), $\kappa^2 (= 2r_0^{-2})$ and its large value implies that, at distances $r \gg r_0$, the magnetic current density ∂^{μ} is negligibly small. Thus, wave equation (2.4) describes the wave characteristics of the vanishingly small magnetic current density at distances $r \gg r_0$ in an approximately flat space-time.

The magnetic current density $\partial^{\mu}(x)$ can be represented as a superposition of plane waves in the form

$$\partial^{\mu}(x) = \left(\frac{1}{2\pi}\right)^4 \int_{-\infty}^{\infty} \partial^{\mu}(k)\exp(ik_{\rho}x^{\rho})\delta(k_{\sigma}k^{\sigma} + \kappa^2)\,d^4k, \tag{2.5}$$

Dirac's influence on unified field theory

where $\partial^\mu(k)$ (i.e., $\partial^\mu(k_1, k_2, k_3, k_0)$) is the Fourier transform of $\partial^\mu(x)$ (i.e., $\partial^\mu(x^1, x^2, x^3, x^4)$) in the k-space or the wave-number space and where the delta function under the integration sign can be expressed as

$$\delta(k^\rho k_\rho + \kappa^2) = \tfrac{1}{2}[\delta(\sqrt{(k_0^2 + \kappa^2)} + k) + \delta(\sqrt{(k_0^2 + \kappa^2)} - k)]. \quad (2.6)$$

In the plane wave

$$\exp(ik_\rho x^\rho) = \exp(ik_0 x^0)\exp(-ik_0 \cdot r), \quad (2.7)$$

k_0 and k_j ($j = 1, 2, 3$) are related to the frequency v and wavelength λ according to $v = ck_0$, $\lambda = |k|^{-1}$, where the velocity of the monochromatic wave is given by $v = \lambda v$. Because of (2.4) and (2.5), the wave number vector k_μ must satisfy the equation

$$k_\rho k^\rho + \kappa^2 = \frac{v^2}{c^2} - \frac{1}{\lambda^2} + \kappa^2 = 0. \quad (2.8)$$

Hence, for the velocity of a wave with frequency v and wavelength λ, we have

$$\frac{v}{c} = \sqrt{(1 - \lambda^2\kappa^2)},$$

or, expressed as a function of the frequency v, we obtain

$$v = \frac{vc}{(v^2 + c^2\kappa^2)}, \quad (2.9)$$

where $\lambda < \kappa^{-1}$. Thus, the magnetic charge current density $\partial^\mu(x)$ is obtained as a superposition of plane waves moving with a velocity less than that of light. However, if the κ^2 terms in the wave equation were negative (i.e., replacing κ by $i\kappa$), then expression (2.9) would yield, for short-range forces, a velocity of propagation greater than light velocity, resulting in tachyons. Thus, the only way to obtain waves moving with the velocity of light is by setting $\kappa = 0$ or $r_0 = \infty$. The latter corresponds to Einstein's version of the unified field theory. The case $r_0 = 0$ yields general relativity in the presence of electromagnetic field.

The conservation of the current $\partial^\mu(x)$ yields the corresponding statement

$$k_\mu \partial^\mu(k) = 0, \quad (2.10)$$

in the wave number space. The definition of $\partial^\mu(x)$ by the divergence

281

equation

$$f^{\mu\nu}_{,\nu} = 4\pi\sigma^\mu, \tag{2.11}$$

in flat space-time, together with (2.5) and (2.10), yield the result

$$f^{\mu\nu}(x) = f^{\mu\nu}_e(x) + \frac{i}{4\pi^3\kappa^2} \int \left[\sigma^\mu(k)k^\nu - \sigma^\nu(k)k^\mu\right]$$
$$\times \exp(ik_\rho x^\rho)\delta(k_\sigma k^\sigma + \kappa^2)\,d^4k, \tag{2.12}$$

where $f^{\mu\nu}_e(x)$ represents an electromagnetic field satisfying the equations

$$f^{\mu\nu}_{e,\nu} = 0, \qquad \left(\nabla^2 - \frac{1}{c^2}\frac{\partial^2}{\partial t^2}\right)f^{\mu\nu}_e = 0. \tag{2.13}$$

Hence, the field $f^{\mu\nu}(x)$, as follows from (2.4) and (2.12), satisfies the wave equation

$$\left(\nabla^2 - \frac{1}{c^2}\frac{\partial^2}{\partial t^2} + \kappa^2\right)f^{\mu\nu}(x) = \kappa^2 f^{\mu\nu}_e. \tag{2.14}$$

Thus, the short-range forces corresponding to $f^{\mu\nu}$ propagate with a velocity less than that of light. The fields $f^{\mu\nu}$, which carry energy and momentum, have a part obeying wave equation (2.14) without the κ^2-term and, therefore, correspond to electromagnetic waves represented by $f^{\mu\nu}_e$. Furthermore, the electromagnetic waves represented by $f^{\mu\nu}_e$ interact with the short-range field $f^{\mu\nu}$ generated by the magnetic charge distribution.

For waves of frequency ν, which propagate with a velocity less than that of light, the wave equation (2.4) becomes

$$(\nabla^2 + k_n^2)\sigma^\mu_n = 0, \tag{2.15}$$

where

$$k_n^2 = \frac{\nu_n^2}{c^2} + \kappa_n^2, \qquad \kappa_n^2 = 2r_{0n}^{-2}. \tag{2.16}$$

According to the fundamental theorems[5] of eigenfunctions and eigenvalues of an equation like (2.15), there exists an infinite system of eigenfunctions

$$\sigma^\mu_1, \sigma^\mu_2, \sigma^\mu_3, \dots, \tag{2.17}$$

whose elements are regular in the interior of a spatial region S and satisfy differential equation (2.15), as well as a homogeneous boundary condition where $\partial_n^\mu = 0$, $n = 1, 2, 3, \ldots$ The corresponding eigenvalues

$$k_1, k_2, k_3, \ldots, \tag{2.18}$$

ordered in an increasing sequence, are infinite in number and increase to infinity, i.e., $v_n \to \infty$, $\kappa_n \to \infty$ (or $r_{0n} \to 0$) as $n \to \infty$. Because of the boundedness of the region S, the eigenfunctions and the corresponding eigenvalues form a discrete spectrum. Eigenfunctions (2.17) tend to zero as $n \to \infty$. It follows from the above analysis that the oscillations of the magnetic current ∂_n^μ in the region S, because of the $\mathrm{Lim}_{n \to \infty} \partial_n^\mu = 0$, tend to zero. In fact, in view of the very large size of k_n as defined by (2.16), all of the currents ∂_n^μ tend to zero with distance from the origin. The velocity of the waves defined by (2.9) for large κ and small wavelength $\lambda \sim (1/\sqrt{2})r_0$ can become very small. Thus, the magnetic current densities ∂_n^μ do not propagate to infinity but are *confined* within the region S. Hence, all the magnetic current densities ∂_n^μ are *confined* to constitute the extended dynamical structure of the elementary particle.

In wave equation (2.14), the field $f^{\mu\nu}$ can be decomposed according to

$$f^{\mu\nu} = f_m^{\mu\nu} + f_e^{\mu\nu}, \tag{2.19}$$

where $f_m^{\mu\nu}$ represents the field due to the magnetic charge distribution at distances large compared to r_0. Substituting from (2.19) in wave equation (2.14) and using electromagnetic wave equation (2.13), we obtain the wave equation

$$\left(\nabla^2 - \frac{1}{c^2} \frac{\partial^2}{\partial t^2} + \kappa^2 \right) f_m^{\mu\nu} = 0, \tag{2.20}$$

for the short-range field $f_m^{\mu\nu}$ produced by the oscillations of the magnetic-charge distribution, and it propagates as a wave with velocity less than that of light. In fact, in view of the very large value of κ, the field $f_m^{\mu\nu}$ is vanishingly small and, therefore, represents the field of *confined magnetic charges*. We have, thus, proven that short-range forces propagate with velocity less than that of light, and the problem posed by Dirac in his letter above does not arise!

3 Stratified distribution of magnetic charges

The nature of the magnetic charge predicted by the present theory is not related to the magnetic monopole (i.e., a particle carrying a net magnetic charge) proposed by Dirac[6] in 1932 and discussed in more detail in his 1948 paper. Dirac's approach was based on establishing a complete symmetry between the existence of electric- and magnetic-current densities and the fact that his theory does not involve gravitation. In the present case, the magnetic charge constitutes the structure of an elementary particle and, therefore, the origin of the density of matter. It emerges as the basic entity, generating a third short-range force in addition to the short-range parts of the electromagnetic and gravitational forces. The distribution of the magnetic charge in an elementary particle produces a short-range force only. Hence, its detection cannot be based on the same premises employed in the detection of an electric charge producing long-range force in addition to the short-range part of the electromagnetic field.

In order to obtain the nature of the magnetic charge distribution, we shall begin by an expansion of plane waves of the form $\exp(\pm ik_n \cdot r)$, which satisfies wave equation (2.15). The well-known plane wave expansion[7] is given by

$$\exp(isk_n \cdot r) = \exp(isk_n r \cos \theta)$$

$$= \sum_{l=0}^{\infty} (2l + 1)(si)^l j_l(k_n r) P_l(\cos \theta), \qquad (3.1)$$

where $s = \pm 1$ and $j_l(k_n \cdot r)$ are spherical Bessel functions and $P_l(\cos \theta)$ represents Legendre polynomials. By using (2.5) and (2.6), we may further express the magnetic-current density eigenfunctions in the form

$$\mathcal{J}_n^\mu(x) = \left(\frac{1}{2\pi}\right)^4 \frac{1}{c} \sum_{s=-1,1} \sum_{l=0}^{\infty} (2l + 1)(si)^l \int_0^\infty dk_n k_n^2 \int_0^{2\pi} d\phi \int_0^\pi d\theta \sin \theta$$

$$\times \int_0^\infty dv_n \mathcal{J}_n^\nu(sk) \exp(isv_n t) j_l(k_n r) P_l(\cos \theta) \delta\left[k_n^2 - \left(\frac{v_n^2}{c^2} + \kappa^2\right)\right],$$

$$(3.2)$$

where the delta function is defined by (2.6) and where the sum over $s = 1$, $s = -1$ yields a complete expansion including both positive and negative frequencies.

For a time-independent, spherically symmetric (i.e., $l = 0$) magnetic-charge distribution at distances large, compared to r_0, we can write

$$\jmath^0 = \pm \frac{q}{4\pi} \frac{\sin \kappa\beta}{\beta}, \qquad \beta = \pm r, \qquad (3.3)$$

where $\jmath^0$ satisfies the equation

$$\left(\frac{d^2}{d\beta^2} + \frac{2}{\beta} \frac{d}{d\beta} + \kappa^2\right)\jmath^0 = 0.$$

Thus, the density changes sign at the points

$$\kappa r = \pm\pi, \pm 2\pi, \pm 3\pi, \dots, \qquad (3.4)$$

and the spherically symmetric distribution consists of *stratified layers* of magnetic-charge densities with alternating signs and decreasing magnitudes (Fig. 1).

Each term in expansion (3.2) for $l \neq 0$ represents a deviation from the spherically symmetric magnetic-charge distribution. These properties of the magnetic-charge distribution apply, as can be seen from wave equation (2.20), to the short-range field $f_m^{\mu\nu}$.

4 Spectrum of magnetic charges

One common characteristic of the exact solutions discussed in ref. 4 and in an unpublished report[8] is the fact that all of them refer to the regions of zero magnetic-charge density (i.e., magnetically neutral surfaces). We must, therefore, seek all other solutions for which magnetic-charge density vanishes.

The points given by (3.4) describe the neutral surfaces or regions of transition lying between opposite signs of magnetic-charge distribution. In general, the existence of magnetic layers with alternating signs of charge densities in traversing through the magnetically neutral surfaces are to be obtained by setting magnetic-charge density equal to zero in the field equations.

The field equations for spherically symmetric space-time (pertaining to regions of zero magnetic-charge density) to be integrated are given by

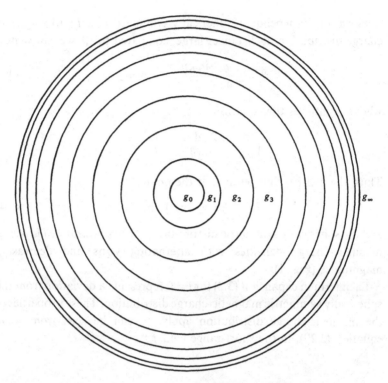

Fig. 1. A spherically symmetric orbiton consists of stratified layers of magnetic charges with alternating signs and decreasing amounts. The sum of the magnetic charges of these infinite stratified layers is zero. The boundaries of the magnetic layers are not sharply defined, but have distributions that are determined from the solutions of equations (4.1)–(4.3). An orbiton is a 'magnet' where each monopole is screened by the adjacent monopole with different magnitude resulting in a short-range force around the orbiton.

(see ref. 8)

$$\frac{d}{dt}\left[St^{-3/2}\sqrt{(1+t^2)}\right]$$

$$=\frac{1}{2}\frac{t^{3/2}}{(1+t^2)^{3/2}}\left[\sqrt{(1+m^2t^2)}-1+(-1)^s tm\right], \quad (4.1)$$

$$\frac{d}{dt}\left[St^{-1/2}\sqrt{(1+t^2)} \right]$$

$$= \frac{1}{2}\frac{t^{3/2}}{(1+t^2)^{3/2}}\left[t - t\sqrt{(1+m^2t^2)} + m(1+t^2) + (-1)^s m \right], \quad (4.2)$$

$$\frac{d}{dt}\left[\frac{dS}{dt}t^{-3/2}\sqrt{(1+t^2)^3} \right] = \frac{1}{2}\frac{mt^{3/2}}{\sqrt{(1+t^2)}}\left[1 - \frac{mt}{\sqrt{(1+m^2t^2)}} \right], \quad (4.3)$$

where the 'eigenvalue' m is defined by

$$m^2 = \frac{l_0^4}{r_0^4} = \frac{g^2}{e^2+g^2}, \quad (4.4)$$

and where

$$t = \frac{r^2}{l_0^2}, \quad r_0^2 = \frac{2G}{c^4}(e^2+g^2), \quad l_0^2 = \frac{2G}{c^4}g\sqrt{(e^2+g^2)}.$$

For the region of zero magnetic-charge density, the field equations, as seen from (4.1)–(4.3), are now decoupled. The integrated form of equations (4.1)–(4.3) is obtained in (VI.42)–(VI.44) of Appendix VI.[9] The solutions for S of equations (4.1)–(4.3) will be represented by the surfaces $S = f_1(t,m)$, $S = f_2(t,m)$, $S = f_3(t,m)$, respectively. Equations (4.1)–(4.3) are not compatible for every S, t, and m. One obvious solution satisfying all three equations (4.1)–(4.3) is given by

$$m = 0, \quad S = 0, \quad (4.5)$$

where t is a finite arbitrary parameter which, together with the function $\exp(u)$ appearing in the definition (IV.1.11) of ref. 8 for S, imply the nonlocalizability of the origin of a particle, i.e., the origin is subject to distribution. From definition (V.3.8) of ref. 8, we have

$$r^2 = l_0^2 t, \quad (4.6)$$

and, therefore, for $m = 0$, we obtain $r = 0$, provided t and the function $\exp(u)$ remain finite. Solution (4.5) is also valid at infinity, where $\exp(\rho)$ tends to infinity and S tends to zero. The latter is no surprise since, in this case, for $l_0^2 = 0$ and $t = \infty$, we should still obtain $r = \infty$.

The surfaces of compatibility for equations (4.1)–(4.3) correspond to the magnetic layers with alternating signs of magnetic charge. From definition (4.4), we see that the magnetic and electric charges are

Behram N. Kursunoglu

related by

$$g^2 = e^2 \frac{m^2}{1 - m^2}, \tag{4.7}$$

where

$$0 \leqq m^2 < 1. \tag{4.8}$$

Relation (4.7) provides, for given spectrum of m-values, a discrete (or 'quantized') relationship between the magnetic-charge values g_n and the electric charge e.

In equations (4.1)–(4.3), the $s = 0$ and $s = 1$ refer to adjacent layers of opposite signs of magnetic charges. Equation (4.3) is independent of the magnetic-charge sign and contains dS/dt, the slope of the function S. Thus, (4.3) is related to the interface between the adjacent layers of opposite signs of magnetic charges (i.e., magnetically neutral surfaces). Hence, we see that the solutions S_+ and S_-, corresponding to the $s = 0$ and $s = 1$ values, respectively (representing transitions in the direction from positive to negative layer and from negative to positive layer, respectively) can be taken together with the dS/dt of (4.3) to posit two sets of algebraic equations to calculate the spectra of t, m, and mG/c^2l_0. The two sets of algebraic equations are given by

$$\text{(i)} \quad f_1^-\left(m, t, \frac{G}{c^2l_0}\right) = f_2^-\left(m, t, \frac{mG}{c^2l_0}\right), \quad \frac{df_1^-}{dt} = \frac{df_3}{dt}, \quad \frac{df_2^-}{dt} = \frac{df_3}{dt}, \tag{4.9}$$

$$\text{(ii)} \quad f_1^+\left(m, t, \frac{mG}{c^2l_0}\right) = f_2^+\left(m, t, \frac{mG}{c^2l_0}\right), \quad \frac{df_1^+}{dt} = \frac{df_3}{dt}, \quad \frac{df_2^+}{dt} = \frac{df_3}{dt}. \tag{4.10}$$

Hence, we see that, in principle, the two sets of algebraic equations (4.9)–(4.10) should yield two different sets of spectral values of $t, m, mG/c^2l_0$ that may represent two different particles. Of course, a corresponding value exists for the function S for each set of spectral value. Furthermore, in view of e^2-dependence of m and l_0, the results apply to both particles and antiparticles. The actual solutions of algebraic equations (4.9) and (4.10), in terms of the integrals as given in ref. 8, require a computer analysis of these equations.

288

5 Complimentarity of spin and structure

The fundamental conservation law $\sum_0^\infty g_n = 0$ for the magnetic charge structure of an elementary particle and the spin and structure relation $\frac{1}{2}\hbar = (1/c)\sum_0^\infty g_n^2$ provide an example for the principle of complimentarity. The measurement of the spin, on the one hand, and the measurement of the magnetic structure, i.e., the observation of the nth partial angular momentum $\frac{1}{2}g_n^2$, on the other hand, constitute complimentary properties of an elementary particle. Thus, if one performs an experiment to measure the spin-angular momentum of a particle, one finds that it has spin. In the same manner, if one performs an experiment to measure magnetic charge compository structure of a particle, one then observes its magnetic charge structure. The observation of any one of these two, spin and structure, properties of an elementary particle will depend on the nature of the experiment performed. At a temperature corresponding to an energy of 10^{21} MeV, the experiment will allow direct observation of the particle's magnetic charge constants. At lower temperatures, we can measure the total spin angular momentum.

In 1975, Dirac called to inform me that he had just seen an article in *Physics Today* describing experiments by Alan Krisch and his co-workers on polarized proton scattering from a polarized proton target. The data indicated that, at large transverse momentum, the cross section is significantly larger when the two spins were parallel than when they are antiparallel. The fundamental result of the experiment was the fact that spin–spin interaction is far from being negligible in large-angle proton–proton elastic scattering at high energies (5–30 GeV).[10] Dirac informed me that Krisch proposed a theoretical model of the proton consisting of three layers of some constituents and that this was somewhat similar to my own results of a particle having the structure presented in the previous section. I immediately called Krisch and pointed out to him that he would have, among others, the problem of the noninvariance of the structure based on a finite number of layers. Furthermore, there was also the problem of proton stability which he could not possibly explain with this model. However, Krisch was more interested in the absence of an explanation based on QCD than the 'success' of his own model.

My own interest in the Krisch experiment was, on the one hand, influenced by Dirac's telephone call and, on the other, by Krisch's proposed layered model, even though the latter may turn out to be quite irrelevant to explaining the experiment. The close relation between spin and structure in this theory implies that the spin-dependent effects and their possible growth with higher energies in the elastic large-angle PP scattering should be predominant.[8] However, Dirac persuaded me that, unless I find a way to calculate differential cross section, the theory will not have an immediate influence on the evolution of physical theories!

6 *Other remarks*

Before leaving Cambridge at the end of 1952, I went to see Dirac to tell him that I was accepted as a postdoctoral research associate at Cornell University in Ithaca, New York, to work with Hans Bethe. He asked me whether I would continue working on my version of the nonsymmetric theory unifying electromagnetic and gravitational fields. He then inquired if I could calculate the fine structure constant from my theory. I asked Dirac, 'Can you calculate the fine structure constant from your theory?' He answered, 'In the future, I might be able to.' My parting remark, as a young and credulous physicist, was, 'In the future, I might be able to calculate it, too.' Many years and many theories later, no one has an answer to that important question.

Dirac always had a great interest in constructing a finite physical theory independent of all renormalization schemes. His contribution to this volume was originally prepared, at my request, for his 80th year Festschrift. The last time I spoke with Dirac on the phone (August 1984), it was about that manuscript. He said he was mailing it to me and that it contained a summary of his feelings on the undesirability of the appearance of infinities in quantum electrodynamics, even if they can be suppressed by the methods of renormalization. He disliked the acceptance of the renormalizability of any field in physics as a principle of physics.

References

1 S. L. Mintz *et al.*, eds., *Fundamental Theories in Physics*, Plenum Press, New York and London (1974).

2 B. Kaufman, *Helv. Phys. Acta Suppl.* **4**, 227 (1956).
3 A. Pais, *The Science and the Life of Albert Einstein*, p. 348, Oxford Univ. Press, Oxford (1982).
4 *Phys. Rev.* **13**, No. 6, 1538 (1976).
5 A. Sommerfeld, *Partial Differential Equations in Physics*, p. 178, Academic Press, New York (1949).
6 P. A. M. Dirac, *Phys. Rev.* **74**, 817 (1948).
7 B. Kursunoglu, *Modern Quantum Theory*, p. 113, W. H. Freeman, San Francisco (1962).
8 B. N. Kursunoglu, 'New Directions in General Theory of Relativity: Uniqueness of Gauge Symmetry,' CTS Report (1985).
9 *Ibid.*
10 A. Krisch *et al.*, Univ. of Michigan Preprint UM HE 85-17 (1985).

Name index

Index

293

Subject index

Index

Index

Magnetic monopoles and the halos of galaxies 174–93
Majorana neutrino 88
Mariacell 6
mathematical relations 110
Maxwell field 96
Meitner–Hupfeld effect 131
memorial to P. A. M. Dirac xi
Merchant Venturers' Technical College 57, 70, 71, 93, 193
method of cords 83
modified minimum subtraction 167
monopole 40, 110
monopole annihilation 188–90
monopole flux 179–81
more scientific ideas 67–224
Mott scattering 240
My association with Professor Dirac 34–6

Nambu–Gatto theories 268
negative energy particle 62
neutron multiplication 80, 83
New York shopping trip 4, 5
NMR tomography 258
Nobel Prize 7, 18
noncommutation 54, 94
nongravitational waves 279
nonrelativistic Schrödinger theory 98
nonrelativistic theory 241
nonsymmetric generalization 277

old quantum mechanics 203–5
one loop 267
Oppenheimer, J. Robert Memorial Price 53
Orbis Scientiae conference 46, 49, 276
Paul Dirac and Werner Heisenberg – a partnership in science 117–62
Pauli matrices 98
Pauli's wave functions 98
perturbation theory 81, 96, 99
photon-electron scattering 102
physical charge 103
Planck 169ff
Planck's law 95
Playing with equations, the Dirac way 93–116
Poisson bracket 120, 207
positron 36, 62, 101
positron emission tomography (PET) 234
positron theory 103, 104, 106
Prasad–Sommerfeld limit 169
preface xvi

preons 267
pretty mathematics 109
Principia 223, 228
projective geometry 72
proton 39, 104, 289

QED 50
quantisation of the radiation field 38, 96, 217–19
quantized singularities in the electromagnetic field 163
quantum chromodynamics (QCD) 165ff
quantum electrodynamics 39, 72
quantum field theory 64
quantum gravity 273
quantum mechanics 12ff
quantum theory 9ff
quantum theory of dispersion 96
quantum theory of ferromagnetism 99
quantum vacuum polarization effects 166
quarks 167, 170ff

Recollections of Paul Dirac at Florida State University 29–33
relativistic electrodynamics 72
relativistic equation of the electron 36
relativistic quantum theory 125–32
relativistic theory of the electron 97, 220–2
Remembering Paul Dirac 57–65
Reminiscences of Paul Dirac 230–3
renormalizability 290
renormalization 102, 106, 107, 109
renormalization group invariant 168
Riemannian space 104
rigid rotator 212
Rutherford scattering formula 103, 240

S-matrix theory 144ff
St John's College 21, 174, 200
scalar wave equation 97, 170
Schrödinger's cat 249–61
Schrödinger's second equation 61
Schrödinger's wave functions 60
Schrödinger's wave mechanics 122, 211–14ff
self-fractioning centrifuge 78
separation power 76
short-range forces 282ff
Solvary conference (1924) 211
Sommerfeld fine structure formula 98, 103, 131
space reflection 109

296

Index